# Aluminum
## by design

# Aluminum
## by design

Sarah Nichols

essays by

with the assistance of
Elisabeth Agro and Elizabeth Teller

Paola Antonelli
Dennis P. Doordan
Robert Friedel
Penny Sparke
Craig Vogel

Carnegie Museum of Art
Pittsburgh, Pennsylvania

Distributed by
Harry N. Abrams, Inc., Publishers

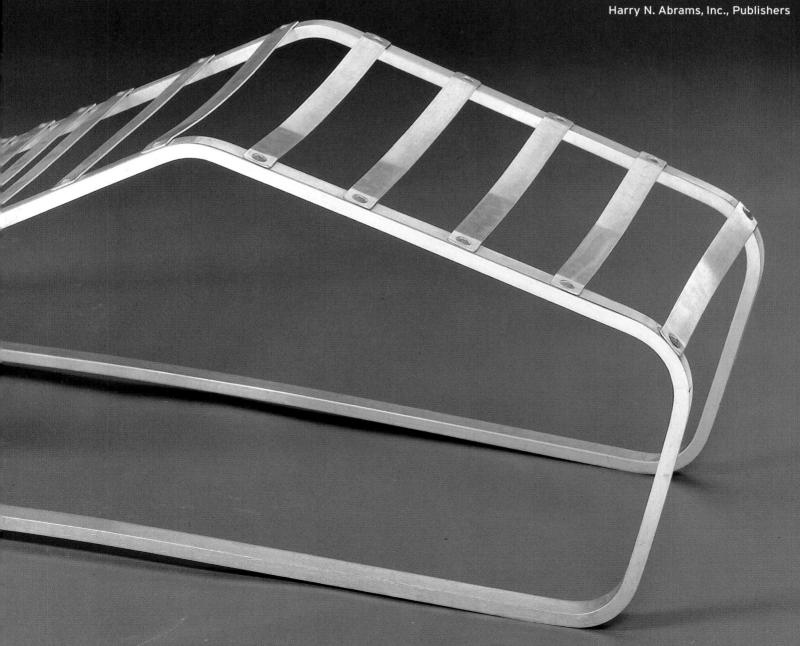

Published on the occasion of the exhibition
*Aluminum by Design: Jewelry to Jets*
organized by Carnegie Museum of Art.

**Carnegie Museum of Art,** Pittsburgh, October 28, 2000 to February 11, 2001
**Cooper-Hewitt, National Design Museum, Smithsonian Institution,** New York, March 20 to July 15, 2001
**The Montreal Museum of Fine Arts,** August 23 to November 4, 2001
**The Wolfsonian–Florida International University,** Miami Beach, December 15, 2001 to April 7, 2002
**Cranbrook Art Museum,** Bloomfield Hills, Michigan, June 1 to August 11, 2002
**Design Museum,** London, October 7, 2002 to January 3, 2003

*Aluminum by Design: Jewelry to Jets* is made possible
by the generous sponsorship of **Alcoa***foundation*.

Significant support has also been provided by **Audi of America, Inc.**

Additional major support has been provided by The Roy A. Hunt Foundation,
The Arthur Vining Davis Foundations, and the Commonwealth of Pennsylvania.

Library of Congress Cataloging-in-Publication Data

Nichols, Sarah C.
  Aluminum by design / Sarah Nichols ; with the assistance of Elisabeth
Agro and Elizabeth Teller ; essays by Paola Antonelli... [et al.].
    p. cm.
"Published on the occasion of the exhibition Aluminum by Design: Jewelry
to Jets organized by Carnegie Museum of Art. Carnegie Museum of Art,
Pittsburgh, October 28, 2000 to February 11, 2001; Cooper-Hewitt, National
Design Museum, Smithsonian Institution, New York, March 20 to July 15,
2001; The Montreal Museum of Fine Arts, August 23 to November 4, 2001;
The Wolfsonian–Florida International University, Miami Beach, December 15,
2001 to April 7, 2002; Cranbrook Art Museum, Bloomfield Hills, Michigan,
June 1 to August 11, 2002; Design Museum, London, October 7, 2002 to
January 3, 2003."
Includes bibliographical references and index.
  ISBN 0-88039-039-5 (pbk. : alk. paper) — ISBN 0-8109-6721-9
(hardcover : alk. paper)
  1. Aluminum—Exhibitions. 2. Aluminum construction—Exhibitions. I.
Agro, Elisabeth R. II. Teller, Elizabeth. III. Antonelli, Paola. IV.
Carnegie Museum of Art. V. Title.
  TA401.6.A1 N53 2000
  620.1'86—dc21
                00-009918

Distributed in 2000 by Harry N. Abrams, Incorporated, New York

© 2000 Carnegie Museum of Art, Carnegie Institute. All rights reserved.
No part of this publication may be reproduced without permission.

Published in 2000 by Carnegie Museum of Art,
4400 Forbes Avenue, Pittsburgh, Pennsylvania 15213–4080

Harry N. Abrams, Inc.
100 Fifth Avenue
New York, N.Y. 10011
www.abramsbooks.com

Project Manager: Gillian Belnap
Editor: Gerald W. R. Ward
Designer: Bethany Johns Design, New York
Typesetting: Moveable Type Inc., Toronto
Lithography and Printing: GZD Grafisches Zentrum Drucktechnik,
Ditzingen-Heimerdingen, Germany
Typeset in Minion, Memphis, and Interstate.
The paper is Tecno matt.

Endpapers:
Extruded aluminum sections,
Aluminum Pavilion, Swiss National
Exhibition, Zurich, 1939 (details)

Title page:
**Marcel Breuer,** designer
Embru-Werke AG, manufacturer
Chaise longue, no. 313, designed
1932, manufactured 1934
(cat. 4.14)

Contents

In 1997, Carnegie Museum of Art approached Alcoa Foundation with an intriguing proposal for an exhibition: a multidisciplinary examination of aluminum's contribution to the evolution of design. Through many media, the story of the art and invention of one material would offer a unique perspective on the role of design in shaping the modern age. The educational potential of Carnegie Museum of Art's concept has been realized now in the museum's presentation of *Aluminum by Design: Jewelry to Jets.*

The fusion of education, art, design, and the history of science and technology in a traveling exhibition extends Alcoa Foundation's tradition of supporting educational and cultural endeavors. Since its establishment in 1952, Alcoa Foundation has strengthened efforts around the world that enrich the quality of our lives by inspiring how we see, think, and imagine the future. Aluminum by Design's comprehensive scope furthers Alcoa Foundation's global commitment to advancing education and the arts.

Bolstered by a successful collaboration with Carnegie Museum of Natural History in the creation of the Alcoa Foundation Hall of American Indians, the Foundation was pleased to work with Carnegie Museum of Art on this project. The museum was poised to make a unique contribution in chronicling the development of design in two centuries through the lens of one material.

Alcoa Foundation is proud to sponsor the international presentation of Carnegie Museum of Art's ambitious effort. This project underscores the museum's international work and creative contributions. We are delighted to see the museum's vision presented at other notable venues, and we salute its talented staff who conceptualized this undertaking and brought it to life.

This exhibition traces aluminum's journey from a once-precious metal to an essential component of everyday life, invention, and art. It educates us about the many individuals who have worked to design and create a better future. It also reminds us that art and design enhance our day-to-day lives in the most unexpected ways. Join us in celebrating *Aluminum by Design: Jewelry to Jets* and its worldwide educational reach.

Kathleen W. Buechel
*President, Alcoa Foundation*

Rarely does a subject, a museum, and its locale fit together so concentrically as aluminum, Carnegie Museum of Art, and Pittsburgh. A 19th-century metal that enabled many of the 20th century's greatest advances, aluminum holds promise for the future. Surveying its evolution to date in the hands of designers is the purpose of the exhibition and book *Aluminum by Design: Jewelry to Jets*.

From its inception four years ago, curator Sarah Nichols and I have been challenged and encouraged by Alcoa Foundation to envision and then realize this ambitious project; we gratefully salute the Foundation's leadership and generosity.

Likewise, we cite with thanks the support of Audi of America, Inc.; The Roy A. Hunt Foundation; The Arthur Vining Davis Foundations; the Commonwealth of Pennsylvania; and Perfido Weiskopf Architects. Substantial underwriting from Billiton guarantees the exhibition's presentation in London. The contribution of the European Aluminium Association is also gratefully acknowledged. The museum's Board has been a keen ally of the exhibition and this volume from the earliest days, as have the members of the Women's Committee.

Sarah Nichols devised, nurtured, and shaped *Aluminum by Design*, recognizing for some time that a comprehensive history of the subject was not available. Her curiosity and discernment account for both its wide scope and its impeccable standards. Synthesizing aesthetics and intellect is a special attribute of leading curators such as Sarah. As you will see, she is to be commended and well merits the admiration of the enormous audience for this book and exhibition.

Aluminum has played an inestimable role in the growth and vitality of Pittsburgh and the world for more than a century. Thus, Carnegie Museum of Art is appropriately proud to share this history of the metal's use in design.

Richard Armstrong
*The Henry J. Heinz II Director*

## Acknowledgments

This publication and exhibition was accomplished with the participation of many dedicated individuals both inside and outside the museum. I am grateful to Alcoa Foundation, especially F. Worth Hobbs and Kathleen W. Buechel; their interest and enthusiasm for this project at the planning stages were decisive in its launch. With Kimberly Hamilton, they have been our praiseworthy partners. The Foundation's continued sponsorship has made the exhibition and publication possible. We are also grateful to Audi of America, Inc. for its support. The further generosity of The Roy A. Hunt Foundation, The Arthur Vining Davis Foundations, and the Commonwealth of Pennsylvania is much appreciated. The contribution of Perfido Weiskopf Architects is also gratefully acknowledged.

The exhibition would not have come to fruition without the initiative of Richard Armstrong, The Henry J. Heinz II Director of Carnegie Museum of Art, who, with former assistant director Michael Fahlund and his successor Maureen Rolla, contributed skill and insight at every stage. Richard's visionary belief that an exhibition of this nature is both appropriate and meaningful for an art museum was an ongoing source of encouragement to me. I also thank Milton Fine, chairman, and the members of the Board, who have demonstrated unfailing commitment to this project through their endorsement of the museum's many aluminum acquisitions.

For their substantial contribution to this publication, I am grateful to essayists Paola Antonelli, Dennis P. Doordan, Robert Friedel, Penny Sparke, and Craig Vogel, whose wide range of expertise greatly enhanced this publication. I am also grateful for their participation in two planning meetings in 1998 and 1999, which informed and shaped the exhibition. It has been a particular pleasure for me to learn from them. My thanks also to museum staff members and other individuals who participated in the planning meetings, with appreciation to Phillip Morton, who has provided invaluable assistance throughout the project.

We are indebted to the private collectors, manufacturers, designers, and public institutions who have lent objects, as well as to those who have donated aluminum works to Carnegie Museum of Art; such generosity has made this exhibition possible. Their names appear in the list of lenders and donors. Additionally, thanks to the following individuals who were particularly helpful in facilitating loans and gifts: Jane Adam, Martin Aurand, Caroline Bacon, Gijs Bakker, Boris Bally, John Bartlett, Pollyanna Beeley, Leslie Calmes, Claire Catterall, Diane Charbonneau, Kevin C. Clarke, Jody Cohen, Norman Craig, Helen Drutt English, Arline Fisch,

Diane Galt, Catherine Gendre, Lisa Katherine Graddy, Margaret Grandine, Gretchen Griswold, Jim Habig, Neil Harvey, Kurt Helfrich, Peter Hollinshead, Larry Huttle, Robert Imig, Coco Kim, Peter Kristiansen, Jun Kunai, Marcia Lewis, Anne Madarasz, Serge Mauduit, Douglas McCombs, Patricia Mears, Christopher Monkhouse, Christopher Moore, Maria Nahigian, Yuko Nakatani, Elias Grove Nielsen, Matthew Nixon, Elise Picard, Donald Peirce, Jean-Michel Piguet, Reinhard Pohanka, Rolande Pozo, Karim Rashid, Stefania Ricci, David Ryan, Bernard Salvat, John S. M. Scott, Claudine Scoville, Peyton Skipwith, Kevin Stayton, Barbara Stewart, Jane Talcott, Ali Tayar, Christian Tilatti, Sebastian Thrun, Trudi Wilner Stack, and James M. Winter, Jr.

We thank as well the many individuals who have volunteered their time to assist in the arduous tasks of locating objects, verifying information, and securing photographs: Laurens van den Acker, Jonathan Betts, George Binczewski, Louis Blum, Bryan Burkhart, Eames Demetrios and the Eames Office, Lorry Dudley, Elizabeth Ferer, Ellen Frola, Deborah Hardy, Alan Heller, Madeleine Hoffmann, Balthazar Korab, Nickolas Kotow, Alberto Meda, Andreas Nutz, Albert Pfeiffer, Alexandra Riegel, Klaus-Josef Roßfeldt, Julius Shulman, Edgar Tafel, and Donald Wright. We benefited continuously from the unflagging assistance of Earl Mounts at the Alcoa Technical Center Research Library, Terrance Keenan of the Department of Special Collections, Syracuse University Library, and the staffs at the libraries at the Henry Francis du Pont Winterthur Museum, Delaware, and the Senator John Heinz History Center, Pittsburgh. I thank in particular Maurice Laparra, Ivan Grinberg, and Florence Hachez-Leroy of the Institut pour l'Histoire de l'Aluminium in Paris for their generous help and hospitality.

This project entailed traveling throughout Europe and the United States, visiting museums, dealers, private collections, manufacturers, and aluminum specialists who graciously gave of their time and expertise. I thank David Allan, Catherine Arminjon, Arthur Armour, Thomas Armour, Paul Asenbaum, George Beylerian, Nicholas Brown, Gregg Buchbinder, Bonita Campbell, Joe Clinton, Mark de Ferrière, William Freeman, Harry Greenberger, David Hanks, Hans Peter Held, Torrence M. Hunt, Sr., Inamarie Klein, Françoise Maison, Michael Malley, Lorraine Mansfield, Vicki Mantranga, Emmanuelle Morgan, Crystal Payton, Leland Payton, David Owsley, Keith Read, Judy Seymour, Bonita Sullivan, Elisabeth Vaupel, Jaroslav Vecko, Giorgio Vigna, Rita Wagner, Christopher Wilk, and Christian Witt-Doerring.

Heartfelt thanks and appreciation go to Jean Plateau. Not only has he been a generous lender to the exhibition, but also he has embraced this project enthusiastically, showing his exceptional collection and, with his gracious wife, Evelyn Plateau, generously entertaining me. Making the acquaintance of the Plateaus has been a great pleasure and wonderful learning experience.

The Women's Committee, led by former and current presidents Jean McCullough and Kennedy B. Nelson, respectively, have been active supporters of this exhibition by allocating acquisition funds for aluminum purchases. They also organized the exhibition's elegant opening celebrations; we extend our thanks to the entire Women's Committee and, in particular, event chair Ranny Ferguson and her committee.

This exhibition and publication could not have been realized without the contributions and enthusiasm of the museum staff; everyone has participated in some way. Space does not allow me to thank everyone individually nor elucidate the various and crucial roles they played. I am enormously grateful to the conservation department, headed by William Real, for their tireless work to make the objects look their best; to the education department, which, under Marilyn Russell's leadership, developed a rich variety of programming to amplify the exhibition; to the registrar's office, under Monika Tomko, for monitoring and coordinating the transport of objects; to the exhibitions department, managed by Chris Rauhoff and including the workshop under Frank Pietrusinski, for their exemplary work installing the exhibition; to the marketing department, guided by Rhonda Goldblatt, for bringing the exhibition to the attention of a wide and varied audience; and to the development department, led by Meg Bernard, for helping to put the exhibition on sound financial footing. I further extend gratitude to my curatorial colleagues for their support and encouragement throughout the duration of this project.

Many people outside the museum made invaluable contributions to this project. My thanks go to Bally Design for their exciting exhibition design and to Dan Boyarski and his graduate design seminar at Carnegie Mellon University for producing the intelligent website architecture; to Teri J. Edelstein for her expertise in securing venues for the tour; and to Red House Communications Inc. and Resnicow Schroeder Associates, Inc. for their marketing campaigns.

My sincere thanks go to all the people who worked on this publication: Gillian Belnap, who expertly managed the project; Gerald Ward for his thoughtful and thorough editing; Elissa Curcio for her attentive proofreading; Bethany Johns for the handsome design of the book and the inspired exhibition logo; and Peter Harholdt for his wonderful photographs. I owe them all many thanks for their commitment to this project.

The enthusiastic and unflagging work of the museum's aluminum "team" made this exhibition and publication possible. Their contributions cannot be underestimated. Rachel Layton Elwes, former assistant curator of decorative arts, initiated the research. Her successor, Elisabeth Agro, brought her scholarship to the research and the publication's glossary; with characteristic cheerfulness, Elisabeth also completed many departmental tasks I was unable to attend to. With ingenuity and tenaciousness, Elizabeth Teller, project research assistant, tracked down countless photographs and endless pieces of information; the project has benefited enormously from her attention to myriad details. Wendy Weckerle, departmental assistant, kept the paperwork flowing and all of us organized and on track; and volunteer Sigi Cleland was our willing translator and library gofer. All worked with diligence, intelligence, and great humor.

Finally, I express my gratitude to Phillip M. Johnston, former director of Carnegie Museum of Art, under whose tenure the idea for this exhibition was conceived. As curator of decorative arts, he inspired and encouraged me to explore the uncharted territories of the decorative arts of the 20th century.

Sarah Nichols
*Curator*

Jane Adam

Airstream Inc., Jackson Center, Ohio

Alcoa Automotive, New Kensington, Pa.

Alcoa Inc., Pittsburgh

Alusuisse Technology & Management Ltd., Communications TCA, Neuhausen, Switzerland

Ron Arad & Associates, Ltd., London

Architecture and Design Collection, University Art Museum, University of California, Santa Barbara

The Art Institute of Chicago

Audi of America, Inc., Auburn Hills, Mich.

Austrian Postal Savings Bank, Vienna

John P. Axelrod, Boston

Gijs Bakker

Bang & Olufsen

Pollyanna Beeley

Charles Biddle and Eileen O'Hara

Biomega, Hellerup, Denmark

The Birkenhead Collection

Brooklyn Museum of Art

Mrs. Louise Buck

Dominique Buisson

Elaine Caldwell

Carnegie Mellon University Architectural Archive, Pittsburgh

Casa del Herrero, Montecito, Calif.

Trustees of Cecil Higgins Art Gallery, Bedford, England

Center for Creative Photography, Tucson, Ariz.

Centre Georges Pompidou, Paris, Musée National d'Art Moderne/Centre de Création Industrielle MNAM/CCI

Cooper-Hewitt, National Design Museum, Smithsonian Institution, New York

DaimlerChrysler Corporation, Auburn Hills, Mich.

Deutsches Museum, Munich

East End Galleries, Pittsburgh

Emeco, Hanover, Pa.

The Fine Arts Society, London

Arline Fisch

Clare Graham

Henry Ford Museum and Greenfield Village, Dearborn, Mich.

Erik and Petra Hesmerg

High Museum of Art, Atlanta

Historical Society of Western Pennsylvania, Pittsburgh

Historiches Museum der Stadt Wien, Vienna

Torrence M. Hunt, Sr.

ICF Group, New York

Institut pour l'Histoire de l'Aluminium, Paris

John Jesse

Kappler Safety Group, Guntersville, Ala.

Kennametal, Inc., Latrobe, Pa.

Mrs. Frederick Kiesler

Knoll, Inc., New York

Kölnisches Stadtmuseum, Cologne

Marcia Lewis

Mrs. George H. Love

Aaron Lown

James and Kathleen Manwaring

Mellon Bank, Pittsburgh

The Minneapolis Institute of Arts, Gift of the Norwest Corporation

Miyake Design Studio, Tokyo and New York

Christopher Monkhouse

The Montreal Museum of Fine Arts / Montreal Museum of Decorative Arts, The Liliane and David M. Stewart Collection

Celia Morrissette and Keith Johnson

Musée de l'Air et de l'Espace, Le Bourget, France

Musée de l'Armée, Paris

Musée des Arts Decoratifs, Paris

Musée des Arts et Métiers—CNAM—Paris

Musée d'Orsay, Paris

Musée Internationale d'Horlogerie, La Chaux-du-Fonds, Switzerland

Musée Lambinet, Versailles

Musée National de Compiègne, France

Museo Salvatore Ferragamo, Florence

Museum für Angewandte Kunst, Vienna

National Air & Space Museum, Smithsonian Institution, Washington, D.C.

National Museum of American History, Smithsonian Institution, Washington, D.C.

Arlen Ness Enterprises

Richard and Jane Nylander

Oberlin College and George J. Binczewski, Oberlin, Ohio

Parallel Design Partnership, New York

Pittsburgh Pirates

Jean Plateau

PowerSki International Corporation, Brea, Calif.

Paco Rabanne, Paris

Karim Rashid, Inc., New York

Paul Reeves

Regional Enterprise Tower, Pittsburgh

retromodern. com

Robot Learning Laboratory, Carnegie Mellon University, Pittsburgh

Rosenborg Castle, Copenhagen

Colin Sayer

Kristine M. Schmidt

SEOS Displays Limited, West Chester, Pa.

Rowena and Everett Smith, Jr.

Sony Electronics Inc., Park Ridge, N.J.

Spalding Sports Worldwide, Inc., Chicopee, Mass.

Mr. and Mrs. DeNean Stafford III

Stedelijk Museum, Amsterdam

Lucy Stewart

Estate of Dudley Talcott

David Tanner

Ali Tayar

Toledo Museum of Art, Ohio

Andrew VanStyn

Vitra Design Museum, Weil am Rhein, Germany

Craig M. Vogel and Dé Dé Greenberg

Stephen P. Webster

Mitchell Wolfson Jr. Collection, The Wolfsonian—Florida International University, Miami Beach

# Aluminum by Design: Jewelry to Jets

Sarah Nichols

The year 2000 was rung in by the traditional dropping of a ball at midnight in Times Square, New York, and at the Royal Observatory in Greenwich, England. The Greenwich ball dates from 1919, when an aluminum ball replaced the earlier wood and leather version. The framework of the New York ball, which was made especially for New Year's celebrations, is a geodesic sphere constructed from aluminum struts.[1] Since its earliest manifestations, aluminum has been considered, at least by some, the metal of a miraculous future just around the corner, so its role in announcing the new year and, in this case, arguably the new millennium, is highly appropriate. Although numerous new materials and composites—for example, titanium, graphite, carbon fiber, and kevlar—were developed during the 20th century, the two materials that have had the greatest impact on industry and everyday life over the last hundred years, aluminum and plastic, originated in the 19th. Both will certainly continue to exert an influence in this new century.

In its early history, aluminum faced a problem common to most new materials: finding a market. Celluloid, the first plastic, was at least researched with a market in mind: it was developed in the United States in 1869 in response to the offer of a prize by a billiards supply company for a material that would substitute for ivory in billiard balls.[2] The quest for aluminum, on the other hand, was a far more intellectual pursuit. By the early 19th century, scientists and chemists knew it existed in the earth's crust, although not in a readily accessible form. Their ambition was to pry it from the ground. Once this was achieved, it was up to the entrepreneurs who established the industrial production of aluminum to search out applications and markets for the new material. Along with metallurgists and manufacturers, marketers, designers, and consumers, they helped determine or design the role of aluminum at different moments in its history.

This role has been conditioned by numerous factors. The history of aluminum and its applications is embedded in shifting social, political, and economic contexts. Industrialization, globalization, production and consumption patterns, warfare, new technologies, culture and lifestyle, demographic change, and environmental concerns have all shaped the history of aluminum. As a product of social invention and human ingenuity, a case study of aluminum provides insight into the broader political and economic currents prevalent at certain historical moments. Aluminum,

Figure 1.
**Otto Wagner** (Austrian, 1841–1918)
Air blowers from Austrian Postal
Savings Bank, Vienna, 1904–06

Figures 2–3.
Views of Jean Plateau's collection
of aluminum objects, France, 1999

in turn, has had profound effects on modern life, both embodying and shaping modernity as concept and as lived experience. Many of these dynamics will be explored in the pages that follow and in the accompanying exhibition, which together contribute to the growing literature on aluminum's impact on culture and society.

In spite of the metal's ubiquity in industry and everyday life, until recently the cultural and social impact of aluminum has received scant attention in scholarly circles.[3] In 1991, the Stadtmuseum in Cologne spearheaded a major exhibition and publication on the history and use of the material. Paris is the headquarters of the Institute for the History of Aluminum, founded in 1986 to encourage research on the metal in all the producing countries. One of the Institute's board members, who has been a generous lender to this exhibition, is building a personal collection of aluminum objects that illustrates, among other things, its impact on daily life (figs. 2–3). To date, his collection numbers well over 10,000 objects. In the United States, there is great interest, particularly among a growing group of avid collectors, in household, gift, and table wares produced from the 1930s through the 1950s by companies such as Wendell August Forge (fig. 4; Highlight 12), Everlast (fig. 56), and Kensington Inc. (fig. 141), to name but a few, and by designers such as Russel Wright (Highlight 18), Lurelle Guild (Highlight 19), and Arthur Armour (fig. 5). Scholars are researching the business histories of these designers and firms, identifying the skilled craftsmen they employed, their design sources, and the sociocultural contexts in which they operated.[4]

Recent attention to aluminum reflects a growing interest in the history of materials in general across a broad spectrum, encompassing primary producers, manufacturers, designers, academics, and consumers. Publications like Ezio

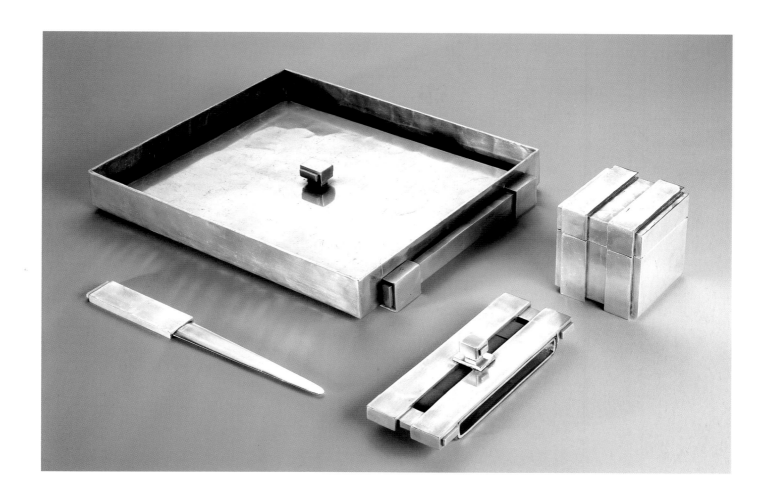

Figure 4.
Wendell August Forge, manufacturer
(American, 1923–present)
Desk set, 1930
aluminum
cat. 3.83

Figure 5.
**Arthur Armour**
(American, 1908–1988)
Butler's tray with zodiac symbols,
ca. 1940
aluminum
cat. 3.84

Manzini's *The Material of Invention* and his seminal work on plastics, exhibitions like *Mutant Materials*, organized by Paola Antonelli at the Museum of Modern Art in New York, and organizations like Material Connexion in New York, established to facilitate the creative exchange of ideas between makers and users of materials, are all part of this process.[5] Consumer awareness is promoted through advertising. The official organizations for the American steel, plastics, and aluminum producers educate consumers through television advertisements on the virtues of each material, so that when offered a choice between a plastic or aluminum container, a steel or aluminum car, materiality may enter into consumers' decisions. Manufacturers, like producers, also advertise their products by promoting their material. For example, publicity for Bang & Olufsen's new BeoLab loudspeaker with anodized aluminum exterior panels provides technical specifications and information on the company's state-of-the-art anodizing plant (Highlight 33).

Environmentalism has also generated interest in materials, particularly aluminum. In varying degrees, consumers, designers, and manufacturers—spurred by activist organizations, governments, or the proverbial bottom line—have developed a commitment to conservation, particularly through recycling.[6] The economics and ethics of recycling are extremely complex, as every stage in a material's life span must be considered. For aluminum, this life-cycle approach takes us from the mining of bauxite to the disposal of the product, as well as the economic and environmental impact of the particular object during its useful life. Just as objects can represent the social and ethical values of an individual or business, so can materials.

Finally, the increasing awareness of aluminum owes much to the popularization of contemporary design—manifest in, among other things, the emergence of the designer as a media superstar and the importance of the company or designer name, label, or logo. (Any parent who has purchased the "wrong" sneakers for a child recognizes the truth of this.) This popularization has encouraged auction houses on both sides of the Atlantic to introduce "design" sales, and accounts for the proliferation of new publications and websites on the topic. Aluminum objects loom large in the history of modern design: in presenting aluminum as shaped by design throughout the 20th century—from its position as a new material to its eventual pervasiveness—one also constructs a general outline of 20th-century design and design issues.

## Key Properties of Aluminum

It is misleading to speak of aluminum as if it were a single material. Aluminum exists in a myriad of forms, each with different qualities and characteristics, and often developed for specific applications. Pure aluminum is relatively soft and ductile, and thus has limited uses. But it can be alloyed with relatively small quantities of other metals and elements to change or expand its properties, while never losing its most important quality—lightness. This "alloyability," along with massive amounts of research into different alloys, propelled aluminum into the pantheon of essential metals. For example, the development of duralumin, an alloy of aluminum and magnesium, copper, and manganese patented in 1910, gave aluminum the strength and hardness necessary for use in airships and airplanes, and eventually led to the fatigue-resistant alloys developed for the aerospace industry. The primary aluminum producers continue to research new alloys and expanded applications, often in conjunction and cooperation with major users such as the aerospace industry and auto manufacturers.

To understand aluminum's preeminence in the 20th century and the often utopian rhetoric associated with it, one needs to know a few basic facts about the material.

- Aluminum is a light-weight metal. It has a low relative density compared to other metals, about one-third that of iron and steel, less than one-third that of copper and its alloys, and one-quarter that of lead and silver.
- Aluminum is ductile and malleable. It can be pulled into thin wire and rolled into thin foil. It can also be cast and extruded into complex shapes, which is particularly important for transportation and construction applications.
- Aluminum is an excellent conductor of heat and electricity (fig. 6).
- Aluminum has a low melting point (660°C) so it is easily recyclable.
- Aluminum is corrosion-resistant, due to a tough film of oxide that forms on its surface when exposed to air. However, this also makes the metal difficult to solder and weld.
- Aluminum can be alloyed with other metals to change and enhance its properties.
- Aluminum can be colored by anodizing (Highlight 33).
- An aluminum surface will accept print (fig. 7).

These properties and qualities—what aluminum can and cannot do—become the basic building blocks in imagining applications and constructing markets for the material.

### Inventing Aluminum: Imitation, Substitution, and Experimentation

The story of aluminum's discovery has been well documented, and is excitingly retold in this volume by Robert Friedel. Past accounts of aluminum's discovery and initial applications tended to have a nationalistic cast. This is largely because, as the *Aluminum News-Letter* stated in 1940, "it was a material that participated to a relatively small extent in international trade as expressed in exports and imports between countries. The production of aluminum in each country is designed primarily to satisfy home requirements."[7] Thus the Swiss considered it a Swiss metal, the Germans a German metal, and so on. Today this picture has totally changed. Aluminum is now a global commodity: the relatively few primary producers are international conglomerates with interests and partnerships all over the world and, with the advent of recycling, aluminum can be produced in areas where the energy requirements of primary production are economically unfeasible.

The story of aluminum in the 19th century belongs almost exclusively to France, from its first production on a useful scale by Henri Sainte-Claire Deville's chemical method in 1854 to the almost simultaneous discovery of the electrolytic process by Charles Martin Hall in America and Paul Héroult in France in 1886. Other producers using the chemical method established aluminum factories in England and the United States, but these were short-lived. However, interest in both Deville's process and aluminum was widespread in international scientific circles. Notices about Deville's discoveries, the properties of the material, and descriptions of aluminum objects, such as medals and watch wheels, began to appear in scientific and popular journals by 1855. In that year, aluminum made its first public appearance at the Paris Exposition.[8] The following year, from August to October, notable scientists such as Michael Faraday, Léon Foucault, and Louis Pasteur visited Deville's laboratory at La Glacière and wrote glowing testimonials to Deville and to aluminum in the visitors' book.[9] But what to do with this new metal? "The recent proposition for the adoption into the arts of the metal aluminium," one journal observed in 1855, "has afforded to the lovers of the marvellous a new subject for speculation. Already its adaptability to a hundred cases is proclaimed…"[10] The article sounded several cautionary notes, the first being that "the difficulty of obtaining [aluminum] is at present a bar to its use." Although this obstacle was eventually overcome, the tendency toward unrealistic expectations and groundless optimism—noted in the same article—became a fixed feature of aluminum "discourse" well into the 20th century.

> Marvels (if we may judge from the eager manner in which they are received by the bulk of mankind) are the natural food for the sustentation [sic] and quickening of many minds. There are few among us who cannot testify to the charm of a really marvellous story. It is indeed quite vexing, while soaring pleasantly in imagination, far from facts and figures, to be brought suddenly, by some friendly monition, into the presence of stern realities, and thereby convinced that we have been merely dreaming…. In short, it is injudicious, whatever custom may say to the contrary, to feed the marvellous vein of craving humanity with fictions based upon supposititious [sic] progress made in the applied sciences.[11]

Compare this somber reality check to J. Barlow's claim in 1856: "It may not be visionary to expect that before this century shall have closed, equally important services in augmenting the comforts of civilized life may be performed by the metal itself."[12]

The key to aluminum's success lay not simply in obtaining quantities sufficient to make a particular application possible at a commercially viable price, but also in finding applications best suited to the material. This is as true today as it was in the 1850s. In the period from about 1856 to 1862, aluminum was still new, exciting, a marvel of science, and a testament to human ability to master Earth (Highlights 1 and 2). It was available in such limited quantities as to make it seem rare, even though its price of $11.75 a pound in 1862 was well below that of silver.[13] Aluminum

Figure 6.
Trivet Manufacturing Co.
(American)
Thermette hot lunch box, ca. 1943
aluminum, leather, plastic,
and fabric
cat. 3.87

Figure 7.
French and Spanish postcards,
ca. 1906
aluminum sheet
cat. 2.58 and 2.59

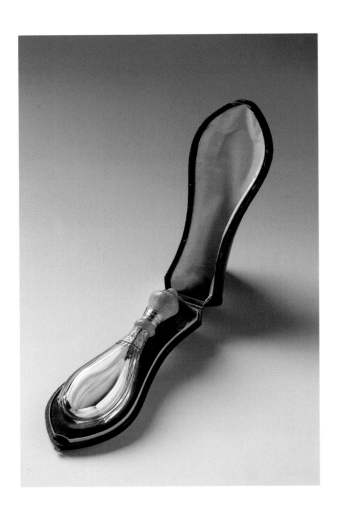

objects from this early period include jewelry and other small keepsakes, often made
in combination with more precious materials, such as the glass perfume bottle with
gold and aluminum cap illustrated here (fig. 8). Other uses were medals and delicate
mechanical and precision parts for watches and scales. The newness and rarity of
aluminum at this time made it highly appropriate for special commissions like the
elaborate cast plaque with a complex mythological scene made at La Glacière in 1856
(fig. 9), and the two ceremonial parade helmets made by Jørgen Balthasar Dalhoff,
the Danish court silversmith, around 1859 for King Frederik VII and his uncle, The
Hereditary Prince Ferdinand (fig. 10). Both must have appreciated the helmets'
light weight in comparison to traditional armor.

    As Deville's process was refined, the price of aluminum fell and its novelty
status waned. Other more practical, sustainable uses for the material needed to be
found based on the characteristics of aluminum itself, rather than on its substitution for other materials. This was recognized at the time. In 1862, at the International
Exhibition in London, where aluminum made "a brave show," a commentator
remarked: "In comparing its colour with that of silver, as a substitute for which it
was once suggested, a marked contrast will be observed, and the conclusion is
evident that it cannot be regarded as a rival to that metal, but must rest its claims
on its own peculiar merit."[14]

    Although aluminum (unlike plastic) was not initially developed to replace
another material, imitation and substitution tend to be the first stage in the history
of any new material. Aluminum was no exception. Sometimes substitution can
be positive: aluminum performed better in some ways than the materials it replaced.
For example, its lightness and resistance to rust and corrosion made it an ideal
material for garden furniture. Between 1925 and 1936, George Steedman designed

Figure 8.
Probably French
Perfume bottle in original case,
ca. 1860
aluminum, glass, gold, and leather
cat. 2.28

Figure 9.
La Glacière (French,
ca. 1855–57)
Plaque, 1856
cast aluminum

Figure 10.
**Jørgen Balthasar Dalhoff**
(Danish, 1800–1890)
Parade helmet for The Hereditary
Prince Ferdinand, ca. 1859
aluminum and other materials
cat. 2.8

and fabricated a set of aluminum garden furniture for his Spanish-style house,
Casa del Herrero, in Montecito, California (fig. 11). The armchair design was inspired
by a pair of Spanish velvet chairs he had purchased in 1923, and the aluminum
seats and backs imitate quilted fabric. An engineer by training, Steedman took up
silversmithing in 1925 as a hobby after retiring from the metal machining business,
and set up a large workshop complete with kilns and a furnace. Clearly not a
man to do things by halves, he contacted leading aircraft manufacturers to learn
how they worked with aluminum.[15] This is an early example of the transfer of technology, seen throughout aluminum's history, in which processes developed for
one application or market are appropriated for another.

   Thoughtful substitution determined by its characteristics is likely to continue
long after a new material establishes its own identity. An example of this is Eero
Saarinen's Tulip Chair, designed in 1956 and still in production today (fig. 12). Saarinen
had wanted to produce the chair in one piece from a single material—plastic.
However, plastics technology at that time was inadequate, and the chair had to be
fabricated from two separate elements: a molded plastic seat shell, and a base cast
from aluminum and coated with plastic to match the seat. The coated aluminum
and the plastic seat worked well together visually, and with its metal base the pedestal could be pared down to resemble the stem of a wine glass, the effect Saarinen

Figure 11.
**George Steedman**
(American, 1871–1940)
Garden furniture at the Casa del
Herrero, Montecito, California,
ca. 1930
aluminum
cat. 2.69

Figure 12.
**Eero Saarinen,** designer
(American, 1910–1961)
Knoll, Inc., manufacturer
(American, 1938–present)
Tulip Chair, designed 1956,
manufactured 1956–present
plastic, cast aluminum
with fused plastic finish,
and upholstery
cat. 4.46

intended. The final product was heavy and thus awkward to maneuver, but if the base had been made of another, heavier metal, it would have been permanently grounded, which was not Saarinen's intention. The Tulip Chair, now a classic that helped revolutionize seating design, was possible only because aluminum could substitute for plastic. And it achieved Saarinen's objective of ridding the domestic interior of a "slum of legs."[16]

Price is another reason to substitute and imitate. Savings occur not only in the cost of the material but also in the methods of fabrication. A good example is the biscuit box made for Carr & Co., based on a pewter box designed by Archibald Knox for Liberty & Co. more than twenty years earlier (figs. 13–14). The aluminum box was less expensive and easier to produce and decorate than the pewter original. Knox's pewter designs for Liberty were themselves examples of one material imitating another: the special composition of the pewter meant that it could be polished to resemble silver.

Aluminum has one major advantage over other metals—its lightness—and this quality has inspired more than a few visionaries. Aerial flight had been imagined since the Renaissance, but in the 19th century, with the discovery of aluminum, suddenly it seemed possible. In 1865, Jules Verne predicted that spaceships would be realized with aluminum. "It is easily wrought, it is very widely distributed, forming

Figure 13.
N. C. J., maker
(British, first half of 20th century)
Carr & Co., retailer
(British, ca. 1920–present)
Biscuit box, ca. 1924
aluminum
cat. 2.54

Figure 14.
**Archibald Knox,** designer
(British, 1864–1933)
Liberty & Co., retailer
(British, 1875–present)
Biscuit box, 1903
pewter
cat. 2.53

Figure 15.
**Gustave Ponton d'Amécourt**
(French, 1825–1888)
Model of a steam helicopter, 1863
aluminum, brass, and copper
cat. 2.9

the basis of most of the rocks, is three times lighter than iron, and seems to have been created with the express purpose of furnishing us with the material for our projectile," he wrote in his novel, *From the Earth to the Moon.*[17] Technology and ideas, the "possible" and the "thinkable"—categories that according to Ezio Manzini shape the material culture that constitutes our everyday environment—were suddenly converging.[18] Futurist thinkers like Verne were not alone in their speculations. Professional metallurgists were soon hailing aluminum as the material of choice for flight. In 1896, aluminum specialist Dr. Joseph W. Richards suggested that "if the problem of aerial navigation is ever satisfactorily solved we may expect to see aluminum largely employed in constructing the machines."[19]

Aluminum was used for the first time in a flying machine in an experimental model of a steam helicopter designed by the Viscount de Ponton d'Amécourt in 1863. The helicopter's boiler and framework were made of aluminum (fig. 15). Unfortunately, the experiment was not a success, and Ponton d'Amécourt, discouraged by serious illness, the indifference of colleagues, and mockery, did not pursue his ideas even though in the same year he cofounded the Society for the Encouragement of Aerial Navigation by Heavier-than-Air Machines.[20] Verne's prophecy and Ponton d'Amécourt's helicopter became reality many years later, once technological advances—including the development of superhard alloys like duralumin—caught up with imagination. But flights of fancy are what fuel invention.

## 1886: A Solution without a Problem

The year 1886 was a watershed for aluminum. This was the year in which Charles Martin Hall in the United States and Paul Héroult in France discovered the electrolytic process for producing aluminum. Their production method, now known as the Hall-Héroult process, is substantially the same as the one in use today. Hall took his patent to a group of Pittsburgh metallurgists and businessmen, who formed the Pittsburgh Reduction Company in 1888. Renamed the Aluminum Company of America in 1907, it is now known as Alcoa. Héroult's patent was taken up by the Aluminium Industrie Aktien Gesellschaft (AIAG) in Neuhausen, Switzerland, which became part of the large aluminum and chemical company Alusuisse-Lonza. Thus two commercially viable aluminum industries came into being at more or less the same time in Europe and America. Like any new business, particularly one employing new and somewhat untested technologies on a large scale, both operations had their teething troubles, technical issues which were eventually resolved.[21] But both also faced a more substantial problem: what to do with their product.

The emergence of these companies must be seen against the backdrop of a changing industrial world. Hall and Héroult's process could not have happened without the development of the electricity industry. Aluminum and electricity supported and benefited each other. The Pittsburgh Reduction Company and AIAG were major users of electricity, and their needs encouraged growth and development in that industry. Improvements in transportation facilitated collection of the necessary raw material and distribution of the finished aluminum and its derivative products. Certainly in Pittsburgh, an entrepreneurial spirit propelled the leading industrialists, who moved in the same social circles and supported each other's products. One of the earliest and largest purchasers of aluminum was Andrew Carnegie's steel company; the addition of small quantities of aluminum during the open-hearth process eliminated pockmarks in steel ingots.

A surviving example of these entrepreneurial crosscurrents is Henry Clay Frick's Pittsburgh home, Clayton. Although not a manifestation of avant-garde style, it was up to the moment technologically. Frick introduced electricity into the house as early as 1888, encouraged by his friend and neighbor George Westinghouse. In 1892, Clayton was substantially altered and expanded by architect Frederick Osterling. As part of the redecorating scheme, Osterling used aluminum leaf to highlight the raised decoration on the ceiling and walls of the breakfast room, and in the dining room, aluminum bosses were sewn onto the leather strapping of the velvet curtains (fig. 16). The rest of the dining-room furniture hardware—hinges and handles on the built-in sideboards—was made of silver, but aluminum, which does not tarnish, was an inspired choice for the curtain trim since silver would have been impossible to clean without marking the leather and velvet. Frick was a friend of the Mellon brothers, who by this date had made significant investments in the Pittsburgh Reduction Company. In the redecoration of the two other formal first-floor rooms at Clayton, the parlor and reception room, undertaken by Cottier & Co. in 1903–04, aluminum leaf was used on the ceilings (fig. 17).[22]

The electrolytic process for producing aluminum is continuous; the smelters cannot be turned off, otherwise the substances in them harden. Producing aluminum was and is a round-the-clock job. Thus, practically overnight aluminum was transformed from a rare metal available in relatively small quantities to an industrial product manufactured in ever-increasing amounts. Price fell as output expanded. In 1886, it was $12.00 a pound in France. In 1892, the U.S. price was 86 cents a pound and had dropped to 36 cents by 1897.[23] By 1892, the Neuhausen plant in Switzerland was producing a thousand pounds of aluminum per day and the Pittsburgh Reduction Company six hundred. Both companies, naturally, were eager to find new and sustainable markets for aluminum.[24]

The potential was clearly there. Trade journals specializing in aluminum prior to 1900 reported on a seemingly endless list of new applications for aluminum in

Figure 16.
**Frederick Osterling**
(American, 1865–1934)
Dining-room curtains, detail,
Clayton, Pittsburgh, 1892

Figure 17.
Cottier & Co. (British and
American, 1869–1915)
Parlor, Clayton, Pittsburgh,
1903–04

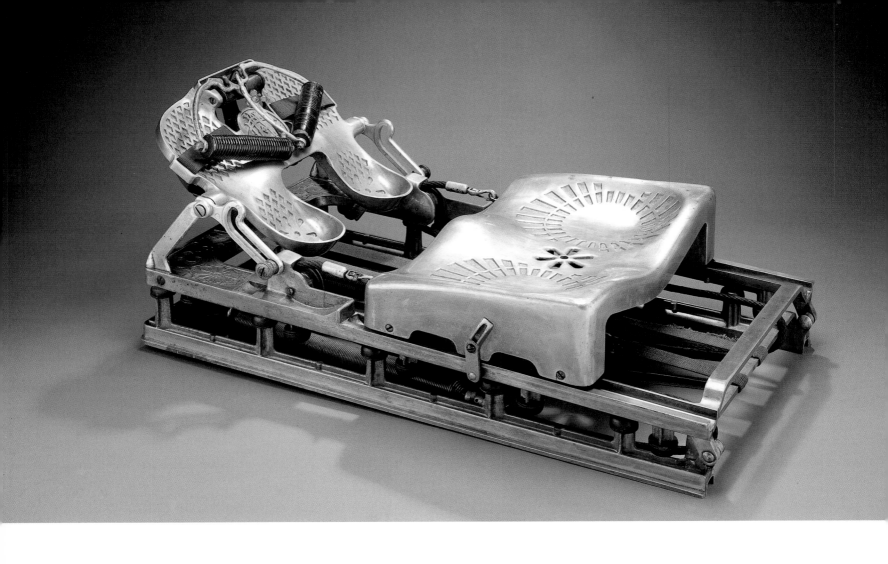

packaging, architecture, transportation, sports equipment (fig. 18), and military hardware—all critical to the 20th-century history of the material in terms of both providing huge markets for producers and dramatically changing the way that people lived. Some new applications, such as electric power transmission lines, capitalized on the metal's unique properties. As an example of its continuing role as a substitute, aluminum foil was used instead of silver for decorating book covers (fig. 21). Other ideas, however, fell prey to the groundless optimism cautioned against in 1855, because of either inadequate technology or public preference for traditional materials. For example, without a suitable alloy, aluminum could not yet meet the requirements for torpedo boats. Neither did the aluminum violin capture the public's imagination, in spite of much rhapsodizing about the metal's sonorous qualities (fig. 19), although the trumpets of many early phonographs were made from aluminum (fig. 20).

But by the turn of the century, aluminum had moved from the rarified atmosphere of Napoleon III's court to the mass market. Novelty items, such as the crab inkwell and others illustrated in trade journal ads, abounded (fig. 22). The aluminum comb market (fig. 23) took off to such an extent that when the Aluminum Association was established in 1901, a special committee on combs was appointed because "the manufacture of combs was of such importance and had so many grievances to adjust." By 1902, *The Aluminum World* reported that more than twenty-five thousand aluminum combs were manufactured daily and "sold all over the world."[25] The concept of novelty

Figure 18.
Health Developing Apparatus Co., Inc.
(American)
Exercise machine, ca. 1905
aluminum, brass, steel, wood,
and cotton rope
cat. 2.41

Figure 19.
Aluminum Musical Instrument
Company (American, active first
half of 20th century)
Violin, ca. 1932
natural lacquered aluminum,
Bakelite, ebony, metal, and wire
cat. 2.52

Figure 20.
Pathé (French, ca. 1895–1928)
Phonograph, ca. 1906–08
aluminum, oak, nickel-plated
steel, and metal
cat. 1.4

(opposite, left)
Figure 21.
**Charles Emile Matthis,** author and
illustrator (French, 1838–unknown)
Jouvet et Cie, printer (French,
active second half of 19th century)
*Les Deux Gaspards*, 1887, and
*L'Expérience du Grand-Papa*, 1888
board, paper, aluminum leaf,
gold leaf, and linen
cat. 2.15

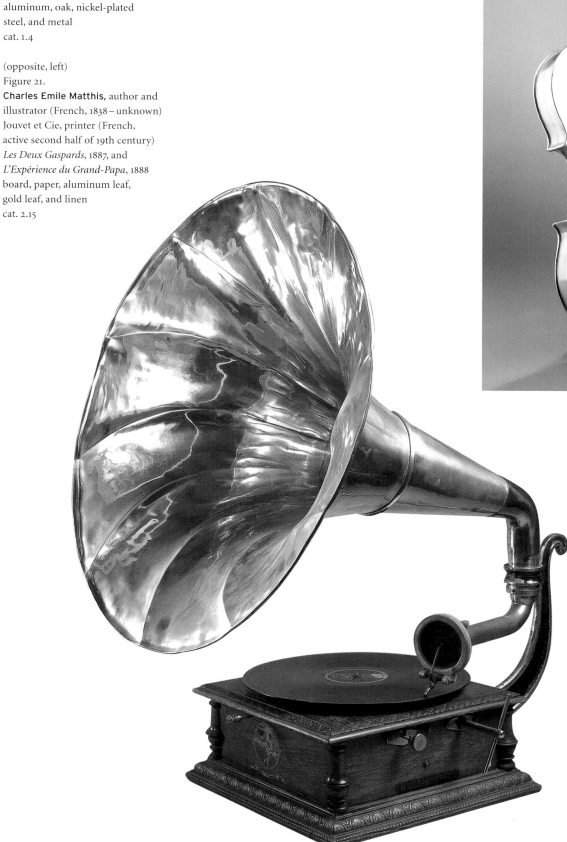

held sway for a long time. As late as 1936, fifty years after Hall and Héroult's discovery, *Aluminium and the Non-Ferrous Review* noted that "novelty remains one of the attractions regarding aluminium goods. To the man in the street, aluminium is still a new metal, and to the ordinary housewife, there is still an air of novelty about a kitchen equipped entirely with aluminium utensils."[26]

Producers of the primary metal, then as now, took the lead in developing new markets and applications, and often became the first manufacturers of new products. They understood aluminum in a way that manufacturers accustomed to working in other metals did not. In these early years, when producers and manufacturers were establishing an identity for aluminum, the effects of poorly made products were not only detrimental but potentially disastrous. The Pittsburgh Reduction Company entered the kettle business as a way to maintain control over quality. After establishing an aluminum product, producers often actively encouraged other manufacturers to expand its market so that they could focus on their primary business, making aluminum.

One application touches on both the creation of a new market and issues of imitation and substitution: all-aluminum furniture. Some of the earliest examples were produced in 1924. In that year, aluminum furniture was shown at the French Salon de l'Aviation, and Alcoa outfitted Mellon Bank's new headquarters in Pittsburgh.[27] That Alcoa should be given the opportunity to enter the furniture field with this large commission (100 desks, 300 chairs, and an assortment of other items) was hardly surprising since Richard Beatty Mellon was a past president of the company and a major shareholder. Metal office furniture was nothing new: Frank Lloyd Wright had designed steel furniture for the Larkin Building in 1904. Resilient and fire-resistant, metal furniture was extremely practical in an office environment. But in its first furniture commission, Alcoa did not intend to make an overtly modernist statement. Perhaps in deference to the conservative atmosphere of Mellon Bank, the company turned to traditional sources for inspiration, particularly for the executive offices; many of the resulting pieces resembled wood furniture in either design or finish, for example, a small table made of parts cast in imitation of turned or carved wooden elements and a coatrack painted in a false wood-grain finish (fig. 25).

Alcoa continued to produce furniture under a variety of brand names, including Alcraft Aluminum. By 1929, the company's Buffalo, New York, factory manufactured a full line of office chairs in about twenty-five different styles at a rate of around three thousand a month. Clients included banks, offices, colleges, hospitals, railroads, hotels, clubs, and restaurants.[28] Two Alcraft catalogues, probably published between 1926 and 1929 and entitled *Distinctive Chairs of Aluminum*, illustrate pared-down but conservative designs set against stylized, geometric, skyscraper-inspired backgrounds (fig. 24). Both catalogues describe the five standard colored finishes available —walnut, mahogany, oak, and two types of green—but the later one also mentions the "beautiful satiny silver" natural finish which "will harmonize extremely well with all modernistic design and coloring."[29] Alcraft was obviously responding to the popular taste of the time since by 1930 metal simulating wood was considered "a mongrel form, dishonest and unbeautiful" among style critics.[30]

Figure 22.
Reymond & Gottlob advertisement for aluminum novelties
*The Aluminum World* 6
(November 1899): 21

Figure 23.
Viko, manufacturer (American, first quarter of 20th century)
Set of combs in original box, ca. 1900
aluminum and cardboard
cat. 2.50

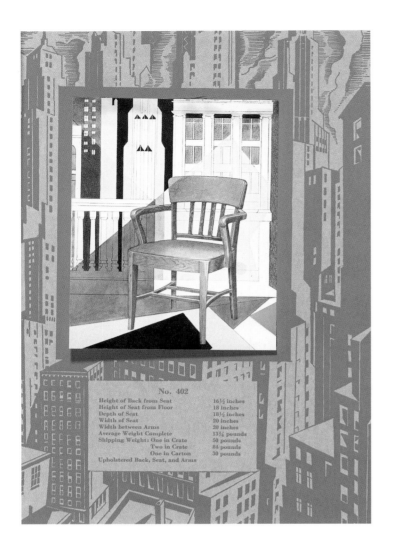

A detailed treatment of aluminum's use in architecture is presented in Dennis Doordan's essay. Here I will simply mention two examples. A very early proposal for aluminum in architecture is illustrated in the 1891 maquette of the facade—including a veranda—of the Théâtre Porte Saint-Martin in Paris (fig. 26). The theater's original stone structure was built in 1873. In 1887, amid concerns about fire safety, a balcony was added. The aluminum maquette may relate to these alterations—certainly metal is a practical building material if fire is an issue.[31] More substantial examples that clearly demonstrate the use of aluminum as a new material in its own right are found in the work of Viennese architect Otto Wagner (Highlight 6). The aluminum furniture, fixtures, and fittings of Wagner's Die Zeit building (1902) and Postal Savings Bank (1904–06; fig. 1), both built in Vienna, established aluminum as a modern, 20th-century material. Aluminum came to epitomize the high-speed, streamlined world that evolved in the 1920s and '30s, and designers and architects employed it in this spirit in the years between the wars.

### The Modernist Ideal

Aluminum production increased dramatically in both Europe and America during World War I. Soldiers on both sides of the conflict carried aluminum canteens by the millions, as well as other aluminum equipment. As George Vits, president of the Aluminum Goods Manufacturing Company, stated, "Not only is the aluminum canteen exceptionally light in weight and rust proof as well, but it offers greater

Figure 24.
"Chair 402," from *Distinctive Chairs of Aluminum* (Buffalo, N.Y.: Alcraft Aluminum, ca. 1926)

Figure 25.
Alcoa (American, 1888–present)
Coatrack and table, 1924
cast aluminum and paint
cat. 2.66 and 2.67

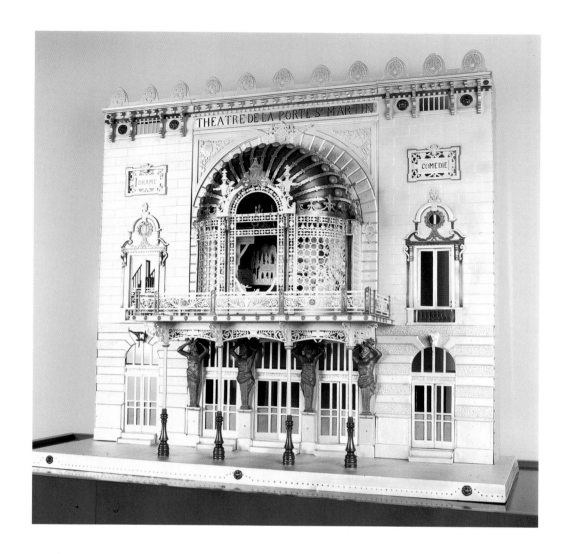

Figure 26.
French
Model of façade of Théâtre Porte
Saint-Martin, Paris, 1891
aluminum, brass, glass, and wood
cat. 2.70

resistance to the mustard gas, which cuts into tin plate and other easily corroded metals."[32] In addition, aluminum is so easy to rework at relatively low temperatures that soldiers reshaped canteens into souvenirs, gifts, and even containers for holy water (fig. 27). Coming out of the war, aluminum producers hoped to maintain wartime production levels, which meant finding new applications. Fortunately, the time was ripe.

Social structures and lifestyles changed rapidly during the interwar years. Advances in technology, increased industrialization, a demographic shift from country to city, the huge loss of life during the war and the worldwide influenza epidemic that followed, the emancipation of women and their changing roles and expectations, and the demise of the servant class all had dramatic impacts on everyday life. The machine age, bolstered by a modernist ideal that embraced the future and the new, not the past, beckoned brightly, at least for some. Aluminum was ideally placed— by virtue of its unique properties, its aesthetic appeal, and the visionary rhetoric long associated with it—to assume the modernist mantle.

This idea of a changing society and the pivotal role that aluminum was to play in this transformation is clearly illustrated in a small publication entitled *L'Aluminium dans le ménage: autrefois, aujourd'hui* (Aluminum in the Household: Yesterday and Today). Published around 1920 by L'Aluminium Français, an organization founded in 1911 to promote French aluminum producers, the book is composed of a series of double-page spreads showing images of life in a French household before and after the arrival of aluminum. Each spread focuses on a different characteristic of the metal; for instance, one compares the hard work

Figure 27.
(left) Unidentified chaser
(French, active ca. 1915)
Rigaworks, manufacturer (Russian,
active beginning of 20th century)
Canteen, manufactured 1913,
chased ca. 1915
chased aluminum
cat. 4.56
(right) French
Holy water container made from
canteen, 1915
aluminum
cat. 4.57

(opposite)
Figure 28.
"Que de temps perdu pour faire
reluire les cuivres" (So much time
wasted polishing copper)
*L'Aluminium dans le ménage:
autrefois, aujourd'hui* (Aluminum
in the Household: Yesterday
and Today) (Paris: L'Aluminium
Français, ca. 1920), 12
paper
cat. 3.90

Figure 29.
"L'Aluminium s'entretient aisément"
(Aluminum is easy to maintain)
*L'Aluminium dans le ménage:
autrefois, aujourd'hui* (Aluminum
in the Household: Yesterday
and Today) (Paris: L'Aluminium
Français, ca. 1920), 13
paper
cat. 3.90

Figure 30.
"L'Aluminium s'emploie partout"
(Aluminum is used everywhere)
*L'Aluminium dans le ménage:
autrefois, aujourd'hui* (Aluminum
in the Household: Yesterday
and Today) (Paris: L'Aluminium
Français, ca. 1920), 18
paper
cat. 3.90

necessary to polish copper with the easily cleaned aluminum (figs. 28–29). The contrast between the two interiors (old-fashioned versus contemporary), the costumes (traditional versus modern), the overall color schemes (dark versus light), and the characters (two servants versus a mother and daughter) clearly positions aluminum as the material of the modern, up-to-date household. The final page, captioned "aluminum is used everywhere," shows a French family sitting down to a meal with an aluminum table setting; an aluminum bread basket and wine cooler are prominently displayed in the foreground. A vignette of images at the top takes aluminum beyond the home, pinpointing its significance in a variety of workplaces, in wiring, and in transportation (fig. 30).

Aluminum had been used in cars, bicycles, airships, and boats in the last years of the 19th century. The interwar period saw huge developments in all forms of transportation, and aluminum played a significant role. *Good Furniture and Decoration* explained this in 1930 in an article on aluminum as a decorative metal: "This is an era of speed and lightness, a period in which there is a clear, concise expression of the essential. To some these feelings can be best portrayed in terms of the white metals, whose history is but comparatively recent. Aluminum is one of the most extensively used white metals of today."[33] The use of aluminum in transportation reduced vehicle weights, which meant not only greater speed but lower operating costs. Aluminum became the material of choice for roadsters and racing cars, milk trucks and motorbikes. In France, Marcel Guiguet developed a motorcycle that made early use of cast aluminum chassis parts, although his vision had not yet been matched by the technology available for either aluminum alloys or engine

Que de temps perdu
pour faire reluire les cuivres.

L'ALUMINIUM
S'ENTRETIENT AISÉMENT.

L'ALUMINIUM S'EMPLOIE PARTOUT.

development (fig. 32).[34] Jaguar Cars, one of the great automotive names, evolved out of the Swallow Sidecars company founded in 1921. Swallow made stylish, zeppelin-shaped aluminum sidecars for motorbikes, enabling a passenger to ride in relative comfort without weighing down the vehicle.[35]

By the mid-1930s, aluminum was used extensively in new trains and buses (fig. 31). By 1939, ten all-aluminum streamlined trains were operating in the United States and two "up-to-the-minute aluminum flyers" were under construction (Highlight 15). In 1935, Greyhound unveiled a fleet of buses manufactured by General Motors Truck Corp., which had "developed an entirely new design to take full advantage of the strength and weight characteristics of the aluminum alloys that were used throughout as the principal materials of construction. The result is a bus with more room and comfort for a larger number of passengers, more baggage space, yet weighing two tons less than the ordinary type of equipment."[36] Similar advances in transportation occurred in Europe and America, epitomized by a French advertisement for aluminum showcased at various salons or exhibitions in 1932, a German advertisement for an aluminum company from 1936, and a 1939 American catalogue for aluminum ladders (figs. 34–36). All three, even the ladders catalogue, depict planes, boats, cars, and trains, clearly highlighting aluminum's status in the modern world. The streamlined nature of transportation design is evident in the American example but also, and at a relatively early date, in the French ad.

Aluminum's structural use in transportation was complemented by its application to interior fixtures and fittings, particularly for trains, ships, and airplanes. The painted cast reliefs for the Santa Fe Railroad, designed by Paul Cret (who had already designed the interiors of the Burlington Zephyr train) and John F. Harbeson, were made of aluminum in order to minimize their weight (fig. 33). The French ocean liner *Normandie*, launched in 1935, included twelve completely metallic cabins with walls, floors, ceilings, doors, and furnishings made of aluminum sheet decorated with stainless, lacquered, and varnished steel (fig. 37).[37] Aluminum seats gradually replaced wicker and rattan chairs in airplanes and airships.

Following the lead of Alcoa and the French aviation industry in 1924, noted designers on both sides of the Atlantic began to produce furniture using aluminum without disguising the material. Metal furniture flourished in the late 1920s and '30s, in the office and beyond. In 1927, the French journal *Art et Décoration* speculated that since metal had progressively replaced wood in furniture for railroad cars and automobiles, perhaps it would do the same for domestic furniture.[38] Architects, designers, and decorators in France heeded the call,

Figure 31.
**Lurelle Guild** (American, 1898–1986)
Drawing of "Streamlined Bus" for
the "Aluminized America" show at
Marshall Field's department store,
Chicago, 1942

Figure 32.
Marcel Guiguet et Cie
(French, 1929–ca. 1940)
MGC N3BR, 1932
aluminum, steel, and other materials

Figure 33.
**Paul Cret**, designer
(American, b. France, 1876–1945)
**John F. Harbeson**, designer
(American, 1888–1986)
Wall ornaments from Santa Fe
Railroad dining car, ca. 1940
aluminum and paint
cat. 3.32 and 3.33

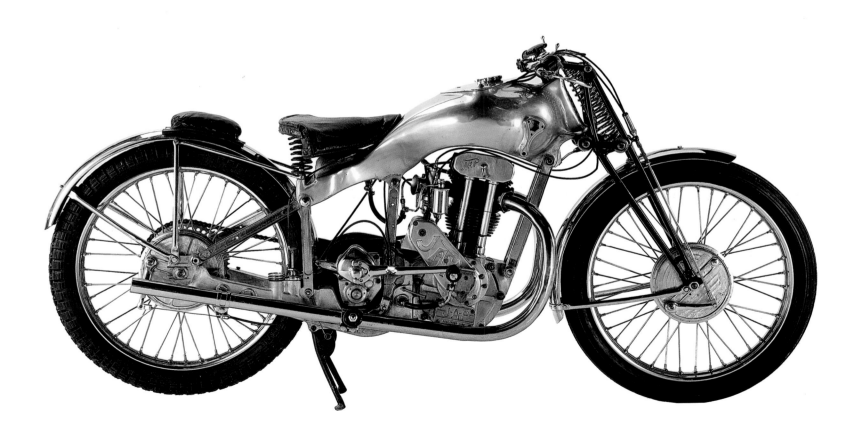

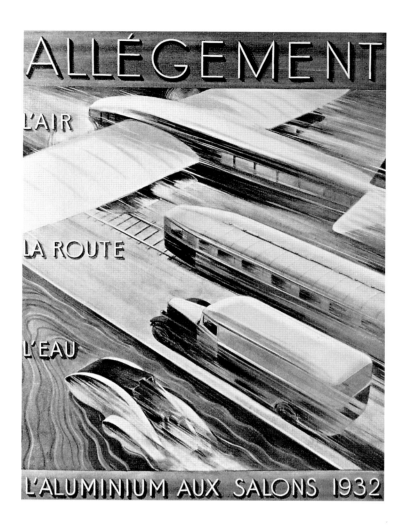

Figure 34.
Advertisement for L'Aluminium
aux Salons 1932: "Allégement: L'Air,
La Route, L'Eau" (Lightness: in
the air, on the road, in the water)
*Revue de l'Aluminium* 52 (1932): 1909

Figure 35.
The Aluminum Ladder Company
(American, active late 1930s)
Cover of *Aluminum Ladders for Every
Requirement* 3 (October 1, 1939)
paper
cat. 3.92

Figure 36.
Vereinigte Leichtmetall-Werke GmbH
advertisement
*Aluminium* (June 1936): XVII/6

*Fig. 22.* — L'oxydation anodique (procédé Bengough) a servi de couche d'apprêt avant application de deux couches de laque pour tous les éléments de construction de deux cabines des six appartements en alliage léger du paquebot « Normandie ». (Décorateur : M' *Klotz* - Construction : C'⁰ *Nord et Alpes*.) (Photo *Marius Gravot*.)

Essai de la Maison Louis Sognot et Charlotte Alix pour une pièce de repos dans une habitation coloniale.

and metal—particularly aluminum and its alloys—began to appear in French furniture. But not everyone was convinced that metal furniture was the ideal route, including the influential British design critic John Gloag, who wrote disparagingly in 1929 that "the designer may devise an interior in which chairs of shining aluminum are an essential part of the composition; but in such schemes human beings appear intrusive, there is no sympathy between them and the setting."[39] With few exceptions, the rest of Britain agreed with Gloag. In continental Europe and America, however, the situation was somewhat different. French architect and designer Charlotte Perriand (Highlight 27) wrote a rebuttal to Gloag in which she placed metal furniture at the forefront of the new age. "Metal plays the same part in furniture as cement has done in architecture," Perriand declared. "IT IS REVOLUTION. The FUTURE will favor materials which best solve the problems propounded by the new man: I understand by the NEW MAN the type of individual who keeps pace with scientific thought, who understands his age and lives it: The Aeroplane, the Ocean Liner, and the Motor are at his service."[40] French furniture designers used aluminum primarily for prestigious or limited commissions, and consequently aluminum furniture in France never reached a mass market as it did in America. For example, in 1927, René Herbst used aluminum for both the interior and exterior of the luxury clothes shop Siegel, located on a new section of Boulevard Haussmann.[41] In 1930, Robert Mallet-Stevens used aluminum for the counter and tables of a bar on Avenue Wagram in Paris. The following year he designed the house and interiors of the Villa Cavrois in Croix and used aluminum extensively in the furniture, including a dressing table and armchairs produced for the lady's boudoir.[42] Louis Sognot and Charlotte Alix exhibited a complete bedroom suite made of duralumin at the 1930 Salon d'Automne (fig. 38).[43] The suite was designed for use in the French colonies, where the temperature and humidity do not suit wood or traditional construction methods. The domestic aluminum industry actively promoted aluminum for both architecture and furniture in the French colonies.

Other European designers experimented with aluminum furniture. Dutch architect and designer Gerrit Rietveld was passionately interested in designing for the mass market. To cut material and production costs, he experimented with commercially available industrial materials formed by minimal construction techniques. In 1942, Rietveld made two prototype chairs from aluminum sheet. Originally he had intended the chair to be produced using folded fiberboard, but this was

Figure 37.
Interior of the French ocean liner *Normandie*
*Revue de l'Aluminium* 72 (1935): 3003

Figure 38.
**Louis Sognot,** designer
(French, 1892–1970)
**Charlotte Alix,** designer
(French, 1897–1987)
Duralumin bedroom suite at the Salon d'Automne, 1930
*Revue de l'Aluminium* 38 (1930): 1294

unavailable due to wartime shortages.[44] Instead, he adapted the design for aluminum, which, surprisingly, he was able to obtain; legend has it that the aluminum came from the wing of a crashed warplane (fig. 39).

Rietveld had wanted to produce a chair using a single sheet of aluminum. This proved impossible, and the added cone-shaped back legs appear awkward. Still, his innovative design would inspire the American sculptor Scott Burton nearly forty years later (fig. 40). A great admirer of Rietveld's furniture and philosophy, Burton's chair pays homage to Rietveld in the choice of material, the triangulation (albeit dramatically attenuated), and the use of circular holes (although Burton himself says that the latter were "inspired by the punched-out patterns used for girders and machine legs beginning in the late 19th century").[45]

Aluminum furniture was produced for a broader economic spectrum in America than in Europe, and Alcoa continued to supply this mass market. From about 1932, Warren McArthur made tubular aluminum furniture for corporate and institutional markets, although he was also taken up by Hollywood, where his aluminum furniture graced the homes of Marlene Dietrich, Ramon Navarro, and Norma Shearer (Highlight 13).[46] Several American designers, including Walter von Nessen, Frederick Kiesler, and Donald Deskey, produced furniture made completely or partly of aluminum. Kiesler designed a nesting coffee table wonderfully organic in shape (fig. 171), and von Nessen, who used aluminum in his lighting fixtures and lamps, designed several geometric, Cubist-inspired tables (fig. 42). Neither Kiesler's nor von Nessen's tables were produced in great numbers, although the Kiesler table has recently been reissued. Donald Deskey used aluminum extensively in his designs for furniture, lighting, and interiors over a number of years. Deskey was interested in many modern industrial materials, including Bakelite, linoleum, and cork, and employed them, often in combination, in both furniture and interior architecture (fig. 43). In the 1920s, Deskey was retained by Reynolds Metals Company in Richmond, Virginia, to find applications for their aluminum foil, initially

developed as a wrapper for cigarettes, as it was ideal for retaining freshness and insulating its contents from moisture. In 1931, Deskey won the competition to design the interiors of Radio City Music Hall. He used new materials throughout the building, including aluminum foil decorated with tobacco motifs as the wallpaper for the men's smoking room, known as the Nicotine Room (fig. 41).[47] Deskey continued to use aluminum after the war. His most notable designs, both dating from 1959, are a suite of furniture known as Charak Modern, produced for the Charak Furniture Company of New York, and the New York City streetlight.[48]

Designers and manufacturers are always making choices about materials. Their criteria are numerous, including cost, availability, aesthetics, public perception of the material, appropriateness, and fabrication techniques. This last consideration is particularly relevant in the choice between tubular aluminum or tubular steel in furniture design. Unlike steel, aluminum cannot be welded easily, and different construction techniques can lead to different design interpretations. A comparison between two versions of a chair designed by Frank Lloyd Wright for the Johnson Wax Administration Building in 1937, one in tubular steel and the other in tubular aluminum, clearly illustrates the point. Wright believed that it was "quite impossible to consider the building as one thing and its furnishings another," and his furniture designs for this commission, particularly the three-legged chair and desk, reflect the building's circular geometry. Wright approached Steelcase Inc. of Grand Rapids, Michigan, and Warren McArthur to produce prototypes in steel and aluminum (figs. 44–45).[49] Under McArthur's production process, the aluminum joints were not welded but had to be secured by circular rings that interrupt the flow of the line. Consequently, the steel version presented a much cleaner design. Wright ultimately chose the Steelcase chair, but whether for aesthetic or financial reasons we cannot be sure.

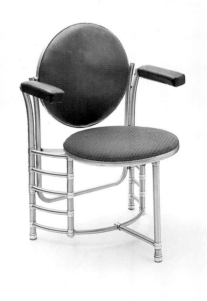

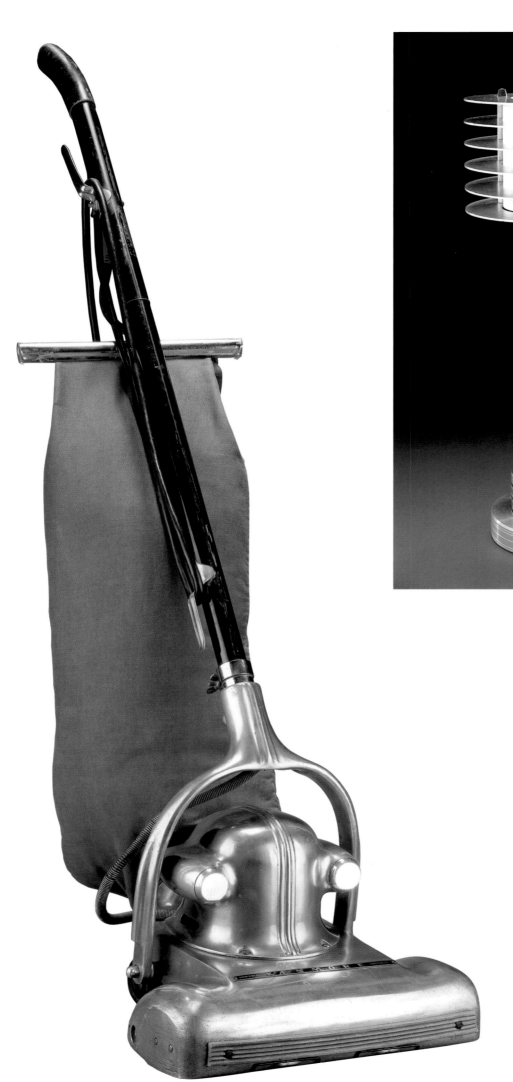

Figure 46.
Sears, Roebuck & Co.
(American, 1893–present)
Imperial Kenmore vacuum
cleaner, 1930
aluminum, metal, iron,
rubber, and cloth
cat. 3.60

Figure 47.
Attributed to **Walter von Nessen**,
designer (American, b. Germany,
1899–1943)
Pattyn Products Co., manufacturer
(American, active first half
of 20th century)
Table lamp, ca. 1935
aluminum, Bakelite, and glass
cat. 3.15

Aluminum was not taken up only by noted designers and architects, or simply used for furniture. The production of aluminum household and consumer goods flourished during the interwar period, as detailed in Penny Sparke's essay. This was curtailed somewhat at the beginning of the Depression, but by the mid-1930s a more optimistic attitude prevailed. As noted industrial designer Henry Dreyfuss wrote:

> …in the coming era of business revival, which will be the greatest period of
> re-design the world has ever known, aluminum will play a large and significant
> part. It will be the basic material of thousands of products. The immediate
> future of industry and science will witness a sharp and decisive battle against
> more weight. The public, its buying power revivified, will demand beauty of
> form, new efficiency of performance and durability.[50]

Aluminum objects invaded the workplace and the home on both sides of the Atlantic. From kettles to vacuum cleaners, aluminum was the ideal material where lightness was sought (fig. 46). Packaging benefited from aluminum's light weight but also from its protective and hygienic qualities (fig. 48). Its reflective qualities were ideal for lighting fixtures (fig. 47). Aluminum conducts heat well, so it was suitable for electric lunch boxes and coffee percolators. Its metallic qualities fit the modern aesthetic, so it was used for decorative or gift wares (fig. 49).

The onset of World War II put an end to—or at least postponed—Dreyfuss's great period of redesign. However, industrial designers had come of age during the interwar years—and had discovered aluminum. More importantly, the aluminum companies had discovered industrial designers and, somewhat out of necessity, turned to them with renewed enthusiasm after the war.

## Conflict and Competition: World War II and Beyond

On March 11, 1941, the American National Defense Commission advised the membership of the Aluminum Wares Association that there would be no more aluminum for cooking utensils. All aluminum was needed for the Allied war effort to manufacture airplanes, motor vehicles, and ships. None could be spared for housewares. In June of that year, the *Aluminum News-Letter* was printed for the first time using nonaluminum ink. This saved twenty pounds of aluminum each month for the defense of the nation. Although hardly a major contribution to the war effort, this gesture of good faith did elevate the publication into the company of "that aluminum window and that aluminum camera tripod and those aluminum streamlined trains that aren't being made temporarily until the emergency is over."[51]

This was a complex period for the American aluminum industry. Alcoa had successfully contended with antitrust suits brought by the government since World War I, but this was to change after World War II.[52] The world had been gearing up for conflict long before war was declared. Part of this preparation involved increasing aluminum production, particularly in Germany in the late 1930s, when its aluminum capacity outstripped that of the United States. The unprecedented increase in wartime demand (U.S. production of aluminum in 1938 was 143,000 short tons and in 1944, 776,000 short tons), and Alcoa's inability to meet it, were the main factors in breaking the company's monopoly. To satisfy demand, the U.S. government financed the rapid building of new aluminum plants. After the war, most of these plants were sold to two new primary producers, Kaiser Aluminum and Chemical Corporation and Reynolds Metals Company. Thus, by 1946 three companies dominated the American market—Alcoa, Reynolds, and Kaiser. *Fortune* magazine outlined the challenges facing the industry in a cover story in May 1946:

> No other major basic industry in the U.S. mushroomed like aluminum during
> World War II. Looking forward, no other basic material has better possibilities
> for expanded peacetime use. But aluminum, after soaring on the wings of
> U.S. and allied air power, is now back to the earth from which it took flight.
> New competition—both within the aluminum industry and between aluminum
> and other materials—has brought new problems and challenges on the road
> to new markets.[53]

These challenges had been anticipated by the aluminum companies as early as January 1942, when Alcoa began to ask the question "how can we maintain employment when the war is over?"[54] Its answer: "imagineering," defined as letting the imagination soar and then engineering it back to Earth. Reminiscent of "feeding the marvellous vein of craving humanity" as described in 1855, or Manzini's "thinkable" and "possible," the sensibility behind Alcoa's "new" concept is evident in the Bohn Aluminum & Brass Corporation advertisements that appeared in *Fortune* in the mid-1940s showing futuristic cars, trains, boats, and planes (fig. 50). In 1944, *Interiors* published an article entitled "Aluminum—A Basic Material of Tomorrow,"

Figure 50.
Bohn Aluminum & Brass
Corporation advertisement
*Fortune* (September 1946): 176

Figure 51.
Bloomingdale's advertisement
*House Furnishing Review* 103, no. 3
(September 1945): 108

Figure 52.
Reynolds Metals Company
advertisement
*Better Homes & Gardens*
(April 1949): 249

Figure 53.
Cartoon depicting recycling cooking
utensils for the war effort
*Punch* (October 1940): 2

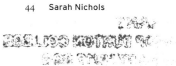

"We're just a frying-pan short on this one."

which promoted the metal to the designers who would shape its practical future: "Aluminum, due to tremendous advances stimulated by the war, will be available in such quantities and at so low a cost that it must now be considered, along with steel, as a basic material upon which the structure of our industrial world will be built."[55]

Other countries faced similar difficulties at war's end: a possible glut in aluminum production, the need to create peacetime employment for a skilled wartime workforce, and the equally urgent need to rebuild and restock. In May 1945, Sir Stafford Cripps, the British government minister responsible for aircraft production, announced an order for fifty thousand temporary aluminum houses. "The construction of these new type houses," *Time* declared, "will keep Britain's light alloys industries, its war expanded plane factories and its skilled army of aluminum workers busily occupied."[56]

As it happened, there was no immediate crisis of overproduction in the United States when the war ended. Rather, the country was ready and eager to replenish, renovate, and rebuild, using aluminum for everything from the consumer goods that people had forgone and recycled (fig. 53) during the war to the new homes needed for returning GIs. Aluminum companies quickly shifted gears. A 1945 advertisement for the housewares department at Bloomingdale's compared an aluminum skillet to a Cézanne painting, a Sèvres vase, and an Iranian rug; the skillet won hands down (fig. 51). A Reynolds advertisement from *Better Homes & Gardens* in 1949 showed an old town being rejuvenated with aluminum (fig. 52). Moreover, with the Korean crisis and the Cold War looming on the horizon, the government was prepared to stockpile. Consequently, the crunch was forestalled until the early 1950s. Then, competition between the various primary aluminum producers and between aluminum and other materials was initiated with renewed vigor and ingenuity, and the three American aluminum companies were poised to exploit design as a promotion and sales technique.

Promotion of aluminum has been key to its development since it was first introduced, and aluminum producers and manufacturers have displayed their products at world's fairs since 1855. For example, the Griswold Manufacturing Company of Erie, Pennsylvania (fig. 118), exhibited "spoons, kettles, frypans, and saucepans, jugs, coffee and tea pots, made entirely of aluminum" at the 1900 Paris Exposition. The proliferation of national and international exhibitions, trade shows, competitions, and design exhibitions, often under the auspices of

Figure 56.
Everlast Metal Products Corporation
(American, active 1937–51)
Souvenir tray from the New York
World's Fair, 1939
aluminum
cat. 4.7

Figure 54.
American
"Golden Temple of Jehol," souvenir
plaque from the Century of
Progress International Exposition,
Chicago, 1934
stamped and painted aluminum,
acetate, polychromed printed
paper, and plywood
cat. 4.8

Figure 55.
American
Set of playing cards from
the Buffalo Exposition, 1901
aluminum
cat. 4.3

Figure 57.
**Lurelle Guild** (American, 1898–1986)
Alcoa museum, New York, 1935

newly formed organizations like AUDAC (American Union of Decorative Artists and Craftsmen) in the United States and UAM (Union des Artistes Modernes) in France, reached a crescendo in the 1930s in both Europe and America (Highlights 14, 20, and 21). Producers, manufacturers, and designers had a vested interest in promoting aluminum wherever possible through these channels, and took every opportunity to do so. As well as seeing impressive displays on every aspect of aluminum, from the heavy industrial to the domestic, visitors to these fairs and exhibitions could often purchase aluminum souvenirs such as a set of aluminum playing cards from the Buffalo Exposition of 1901 (fig. 55), an Everlast tray from the 1939 New York World's Fair (fig. 56), or a plaque from the 1934 Century of Progress International Exposition in Chicago (fig. 54).

Several American museum exhibitions helped elevate the status of industrial design and designers while bringing materials to the fore. *Design for the Machine*, organized in 1932 by the Philadelphia Museum of Art (then called Pennsylvania Museum of Art), showcased contemporary industrial art, emphasizing the importance of understanding the unique properties of any given material and its most advantageous applications. Aluminum chairs, tables, lamps, venetian blinds, kitchen utensils, and smoking stands figured in the exhibition, and Russel Wright (Highlight 18), one of the many designers actively involved in the exhibition, created an aluminum breakfast room.[57]

Alcoa was aware of the power and potential of design, and employed industrial designer Lurelle Guild to enhance the products of subsidiaries such as the Aluminum Cooking Utensil Company, Wear-Ever, and Kensington Inc. from at least 1932 onward (Highlight 19). In his design for Alcoa's aluminum showroom (or museum) at its New York office on Park Avenue, which opened in late 1935, Guild mixed domestic and industrial objects and components in a formal setting that seemed to pay homage to the 1934 exhibition *Machine Art*, curated by Philip Johnson for the Museum of Modern Art, New York (fig. 57).[58]

After the war, Alcoa, Reynolds, and Kaiser recognized that one strategy for taking on the "new challenges on the road to new markets" was to harness design in the way predicted by Dreyfuss in 1935. One obvious way to do this was to get the design community, a profession that had exploded in the 1920s and '30s, excited about aluminum. By 1956, each company had an in-house design staff whose mandate was to work with designers outside the industry, providing any assistance necessary to help them understand the properties of aluminum and its many alloys, and encouraging them to find new uses for the material.[59] The design initiatives of all three companies also had a more public face intended to make the consumer comfortable with aluminum. In this regard, Alcoa launched a high-profile advertising campaign, the Forecast Program (Highlight 28). Over a five-year period, the company commissioned some twenty designers to produce objects from aluminum that could then be used in advertisements. The list reads like a veritable who's who of American design, including Paul McCobb (fig. 58), Harley Earl, Elliot Noyes, Alexander Girard, Charles Eames (Highlight 30), Isamu Noguchi (Highlight 29), and Herbert Bayer, and the objects ranged from a rug by Marianne Strengell (fig. 59) to Garrett Eckbo's garden (fig. 60). Reynolds focused on architecture, commissioning a two-volume

Figure 58.
Alcoa Forecast advertisement featuring Paul McCobb's chairs *Newsweek* (November 25, 1957): 114

Figure 59.
**Marianne Strengell**
(American, b. Finland, 1909–1998)
Rug, 1956
aluminum, jute, wool, and viscose
cat. 4.27

Figure 60.
**Julius Shulman,** photographer
(American, b. 1910)
**Garrett Eckbo,** designer
(American, 1910–2000)
Alcoa Forecast garden with aluminum screen walls and trellis, Laurel Canyon, Los Angeles, 1959

illustrated book on aluminum in modern architecture, publishing interviews with contemporary architects about their use of the metal, and initiating an annual architectural award. Kaiser, which was based in California, sponsored the Hall of Aluminum Fame in Tomorrowland at Disneyland, which included a section on "The Product of Tomorrow" showing "items under development that offer great promise of better living with aluminum in the future," as well as a spaceman in an aluminum suit and a large aluminum telescope (fig. 61).[60]

Along with the competition between aluminum companies, the postwar years marked the start of an intense rivalry between aluminum and other materials industries, especially when a large market share was at stake. Craig Vogel discusses some of the products that have been subjected to a "materials war" in his essay. This competition has been particularly acute in the three arenas identified by the primary aluminum companies as their target markets for the 21st century: packaging, transportation, and construction. The battle for the soda and beer can has been won in the United States, where aluminum is now used exclusively rather than steel, although plastic bottles still maintain a strong market share. In other parts of the world, steel is still used for economic reasons. For example, recycling may not be viable and therefore aluminum would not be cost-effective. The major materials battle now involves cars. In recent years, some very expensive aluminum cars have been produced, such as the Jaguar XJ220, often after close collaboration between the aluminum producer (Alcan of Canada in Jaguar's case) and the car manufacturer (Highlight 42). Innovative technology developed for cars like the Audi A8, such as consolidated castings, special high-strength alloys, the space frame, and new methods for joining body parts, should ultimately be used in the production of more modestly priced vehicles. The choice of material is rarely straightforward but rather is based on many criteria. An overriding factor in the equation today, particularly with cars, derives from the environmental considerations of recycling (Highlight 39), minimizing fuel consumption and emissions, and designing for disassembly (Highlight 40). However, these must be accomplished without raising costs or prejudicing public perceptions of safety.

Figure 61.
H. S. Crocker Co. (American)
Postcard for Kaiser Aluminum
Exhibit in Tomorrowland at
Disneyland, ca. 1955
paper
cat. 4.25

Transportation, packaging, and construction—the three main markets for aluminum—have helped shape the 20th century and will continue to affect everyday life in the 21st. Design has played a determining role in each of these sectors. Although not necessarily critical in terms of market share, aluminum is found in a myriad of other design environments, increasing public awareness of the metal's versatility and redefining public perceptions of its potential applications.

### Crossing Boundaries: Variety and Versatility in Design

In the pluralistic and eclectic design practices of the last thirty or forty years, aluminum has played many roles. It continues to be employed in designs that run the gamut from mass-produced items to exclusive one-offs. Its long history as an industrially produced material has bequeathed it a tradition and vocabulary that can be elaborated upon and exploited. Designers who want a "retro" look that recalls the streamlined aesthetic of the 1930s or the organic forms of the 1950s can achieve it with aluminum. Yet the metal still embodies the idea of the future since it plays a role in the aerospace and robotics industries (Highlight 41) and is an integral component of innovative technology in auto manufacturing.

The boundaries that aluminum crosses and the multitude and variety of successful applications attest to its versatility. Designers of both high fashion and practical work clothing have embraced aluminum. Sometimes its presence is overt, at other times concealed, since it can provide both surface and structure. Aluminum's production process, aesthetics, forms, and alloys developed for one industry are appropriated for other totally different design spheres. The metal is as celebrated for its colored surfaces as for its sleek, cool, silvery look. The intensity and range of colors available with the addition of dyes in the anodizing process have been exploited by both artists and manufacturers (Highlight 33). At the other end of the spectrum, the Japanese retailer Muji uses aluminum's clean "metalicness" in a broad range of small-scale aluminum objects that embody the shop aesthetic, in which bright primary colors are conspicuously absent (fig. 189).

Aluminum continues to be used in the many arenas where its properties make it the material of choice, such as stacking and outdoor furniture, although manufacturers and designers are always looking for new and innovative variations on old themes. Manufacturers can tap "celebrities" for new designs, as Knoll did when it asked Frank Gehry to design a chair to add to its list in the important stackable indoor/outdoor chair market. Knoll wanted a chair with a lower retail price than the ubiquitous but expensive aluminum Toledo model by Jorge Pensi (fig. 62), which the company distributes. The result is the FOG chair (fig. 63), inspired by Gehry's architecture. (FOG derives from Gehry's initials.) Visually compelling ribs and creases articulate the aluminum seat and back and provide structural rigidity, which allows the metal to be thin in places and thus optimizes its light weight. The seat and back are separate because a continuous monocoque form required too expensive a mold. Gehry exploited this segmented design to differentiate between

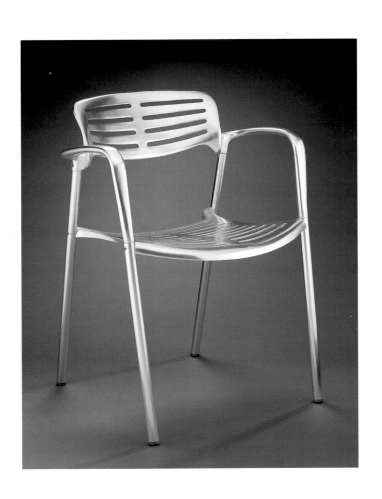

(left)
Figure 62.
**Jorge Pensi,** designer
(Spanish, b. Argentina, 1946)
Amat s.a., manufacturer
(Spanish, 1944–present)
Toledo chair, designed 1986–88,
manufactured 1989–present
epoxy-coated cast aluminum
and aluminum tubing
cat. 5.26

(right)
Figure 63.
**Frank Gehry,** designer
(American, b. Canada, 1929)
Knoll, Inc., manufacturer
(American, 1938–present)
FOG chair, prototype, 1999;
designed 1999, manufactured 2000
aluminum and stainless steel

(below)
Figure 64.
**Ron Arad,** designer (Israeli, b. 1951)
Kartell SpA, manufacturer
(Italian, 1945–present)
FPE (Fantastic Plastic Elastic) chairs,
designed 1997, manufactured
1999–present
plastic and extruded aluminum
cat. 5.37

the two aluminum parts. Carl Magnusson, senior vice president of design at Knoll, rejected plastic for this commission because, as he stated, plastic chairs "suck the life out of a space and leave you feeling diminished." Gehry's architecture provides the cultural context and framework for the FOG chair, and so imparts to it an imposing, powerful, and very physical presence.[61]

Another example of a competitively priced stacking chair suitable for outside use, created by another design name, is Ron Arad's FPE, or Fantastic Plastic Elastic (fig. 64), manufactured by Kartell. The continuous seat and back is made from plastic held in place by double-barreled extruded aluminum tubes bent to form the chair's profile. This tube is cut to leave one short and one long tube that reverse so there is absolutely no waste. The aluminum is treated to negate the metallic effect so that, although silver colored, it resembles plastic in both look and feel. Like Saarinen's Tulip Chair, this is an example of aluminum providing structure to support what is in essence a plastic chair.[62]

Many applications for aluminum that are now popular originated in the 19th century, although they are being pursued with greater vigor today because the design-conscious public is more receptive to unusual materials appearing in unexpected places. In 1898, *Aluminium and Electrolysis* described a dress of woven aluminum produced for a member of Queen Victoria's court as "the first dress ever made containing such a large proportion of aluminium."[63] The time-consuming and labor-intensive weaving technique produced only about half a yard of material a day. In the 1930s, a number of designers and manufacturers, among them Russel Wright, experimented with aluminum foil clothing, but aluminum yarn that could be woven into fabric was not developed until the 1940s and was commercially available only after the war.[64] Aluminum foil, sandwiched between transparent plastic film, was slit into strips as narrow as one-sixty-fourth of an inch, then woven into fabric. The adhesives binding the foil and plastic could be tinted to provide the desired color. This type of metallic yarn was produced under a number of trade names, including Lurex, Reymet, and Metlon, by companies such as Dobeckman, which specialized in unusual and challenging yarns. These yarns were appropriate for bathing suits (fig. 65) and evening and cocktail dresses, as well as upholstery and drapery fabric. Several designers, such as Dorothy Liebes and Marianne Strengell, often working closely with firms like Dobeckman, experimented with aluminum in their textiles. Aluminum is also the practical clothing material of many workplaces, from the butchers' aprons made of aluminum disks resembling chain mail that are worn in the abattoirs of France (Highlight 31), to the aluminum-coated protective clothing of the firefighter and steelworker, to the multilayered spacesuit of the astronaut.

Metallic yarns look and act like fabric. Aluminum squares and disks behave very differently, and the 1960s saw the introduction of such nontraditional materials into the world of high fashion. Paco Rabanne (Highlight 31) produced minidresses of plastic and aluminum in the late 1960s, and his 1999 collection contained a number of aluminum disk dresses. Aluminum is also used by artists and couturiers for accessories, such as Salvatore Ferragamo's sleek and sophisticated bags (fig. 66).

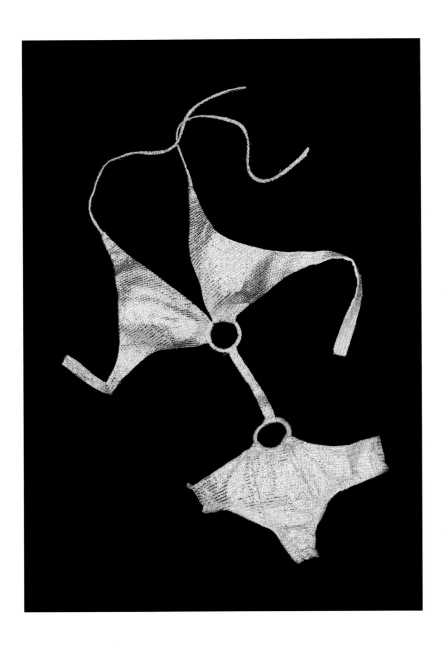

Figure 65.
**Oscar de la Renta** (American,
b. Dominican Republic, 1936)
Bathing suit, ca. 1967
aluminum and synthetic fabric
cat. 5.11

Figure 66.
Salvatore Ferragamo Company
(Italian, 1927–present)
Handbags, 1996–97
anodized aluminum, nickel,
and fabric
cat. 5.15 and 5.16

Figure 67.
**Pollyanna Beeley** (British, b. 1970)
Bodice, 1993
anodized aluminum
cat. 5.10

Pollyanna Beeley, a young British artist whose primary material is anodized aluminum, has produced bags as well as a dramatic bodice (fig. 67).

Interesting dialogues can develop between the various users of aluminum. One example of this is seen in the work of British artist Jane Adam (Highlight 33), who uses aluminum extensively in jewelry and has been a pioneer in experimentation with anodized and textured aluminum. Adam obtains her aluminum from British Industrial Graphics, a company based in Oxfordshire that has used the material for many years in the manufacture of fascia and trim for other industries. Recently, the company was approached by Rover, the car manufacturer, to develop an aluminum trim with a spun-and-turned effect for the dashboard of its new fortieth-anniversary Mini car. Rover wanted something sporty that aesthetically and technically recalled the hand-patterned dashboard of the old Bentley, but without the £5,000 per square meter price tag. Recognizing a potentially large niche market, British Industrial Graphics has produced a wide range of commercially available etched patterns for aluminum trim. The company brought to the mass market an interest in surface quality and texture, something that artists like Jane Adam have been working on for a number of years.[65]

The transfer of materials and technologies, as well as other instances of appropriation from one industry to another, occurs frequently in today's multifaceted, multidimensional design world, and not just with aluminum. Many designers, such as Ron Arad (Highlight 35), Ross Lovegrove, and Alberto Meda, have made use of the honeycombed and superplastic aluminum developed for the aerospace industry in furniture and other objects. Stuart Basseches and Judith Hudson, the founders of Biproduct, use aluminum I-beams adapted from the construction industry in their modular furniture (fig. 68). Werner Schmidt's limited-edition aluminum folding table (figs. 69–70) is a work of precision engineering constructed from aluminum sheets cut by laser with CAD assistance. The sheets are both stable and thin, allowing the eight sections of the table to fold flat to a depth of only 2⅜ inches (6 cm), but the hundreds of hinges are riveted to the sheets by hand in a workshop environment, producing a wonderful symbiosis of high technology and handcraft.

Today, aluminum is equally at home in the aerospace industry and the jewelry artist's studio, and plays a significant role in both. The metal's journey through the 20th century has been varied and not always smooth. Aluminum producers have worked hard to find, develop, and maintain markets for their product. And the quest continues. The metal has been subjected to the usual boom and bust of economic cycles, as well as changing lifestyle and consumption patterns. Aluminum has been different things to different people at different times and will continue to be so, although in this pluralistic age, many roles and vocabularies for the metal can exist harmoniously. One thing is certain: without aluminum, life in the 20th century would have been very different. There is no reason to doubt that aluminum will continue to shape the new millennium in ways equally profound.

## Notes

1. *New York Times*, December 30, 1999, D1. Information kindly provided by Jonathan Betts, National Maritime Museum, London.

2. Jeffrey L. Meikle, *American Plastic: A Cultural History* (New Brunswick, N.J.: Rutgers University Press, 1995), 10.

3. The role of plastics, on the other hand, is well recognized and documented. See Meikle, *American Plastic*; Penny Sparke, ed., *The Plastics Age: From Modernity to Post-Modernity* (London: Victoria and Albert Museum, 1990); Stephen Fenichell, *Plastic: The Making of a Synthetic Century* (New York: HarperBusiness, 1996).

4. Werner Schäfke, Thomas Schleper, and Max Tauch, eds., *Aluminium: Das Metall der Moderne: Gestalt, Gebrauch, Geschichte* (Cologne: Kölnisches Stadtmuseum, 1991). The Institute for the History of Aluminum produces a scholarly journal, organizes symposia and exhibitions, and supports doctoral research; for more information on its activities, contact Institut pour l'Histoire de l'Aluminium, Tour Manhattan, 92087 Paris La Défense Cedex, France. For examples of 1930s–50s wares, see Bonita J. Campbell, *Wendell August Forge: Seventy Five Years of Artistry in Metal* (Upland, Calif.: Dragonflyer Press, 1999); Bonita J. Campbell, *Depression Silver: Machine Age Craft and Design in Aluminum* (Northridge: California State University, School of the Arts, 1995).

5. Ezio Manzini, *The Material of Invention* (Cambridge, Mass.: MIT Press, 1989); Paola Antonelli, *Mutant Materials in Contemporary Design* (New York: Museum of Modern Art, 1995); Doug Fitch, "Connecting…with Materials," *Metropolis* (May 1997): 84–87.

6. Victor Papanek, *The Green Imperative: Natural Design for the Real World* (New York: Thames and Hudson, 1995), 29–48.

7. "Wider Use for Aluminum," *Aluminum News-Letter* (January 1940): 1.

8. "The New Metal—Aluminium," *The Crayon* 2 (September 19, 1855): 183; Jean Plateau and Elise Picard, "L'aluminium, métal de l'argile," *La Revue* 25 (Musée des Arts et Métiers, Paris) (December 1998): 14–23.

9. The visitors' book is in the collection of the Royal Ontario Museum, Toronto.

10. "Aluminium, the So-called New Metal," *Newton's London Journal of the Arts and Sciences* (September 1855), reprinted in *Journal of the Franklin Institute* 61 (1856): 27.

11. Ibid.

12. J. Barlow, "On Aluminium," *Proceedings of the Royal Institution* 2 (March 14, 1856): 222.

13. George David Smith, *From Monopoly to Competition: The Transformation of Alcoa, 1888–1986* (Cambridge: Cambridge University Press, 1988), 7, 34.

14. Originally published in *The Record of the Great Exhibition*, published as a supplement to the *Practical Mechanic's Journal*, quoted in "The History of Aluminium in England," *Aluminium and Electrolysis* (December 1896): 75.

15. David Gebhard, "Casa del Herrero, the George F. Steedman House, Montecito, California," *The Magazine Antiques* (August 1986): 280–93; copies of correspondence between Steedman and Douglas Aircraft Company kindly supplied by Casa del Herrero Foundation.

16. Alexander von Vegesack, Peter Dunas, and Mathias Schwartz-Clauss, eds., *100 Masterpieces from the Vitra Design Museum Collection* (Weil am Rhein, Germany: Vitra Design Museum, 1996), 162; Charlotte Fiell and Peter Fiell, *Design of the 20th Century* (Cologne: Taschen, 1999), 622.

17. "1865—Un voyage en littérature: De la terre à la lune—Jules Verne," *Cahiers d'Histoire de l'Aluminium* 5 (Autumn 1989): 34–35.

18. Sparke, *The Plastics Age*, 7.

19. Joseph W. Richards, *Aluminium: Its History, Occurrence, Properties, Metallurgy, and Applications, Including Its Alloys*, 3d ed. (Philadelphia: H. C. Baird & Co., 1896), 478.

20. Pierre Lissarrague, *Premiers envols* (Paris: J. Cuénot, 1982), 93; Joseph Lecornu, *La navigation aérienne: Histoire documentaire et anecdotique*, 3d ed. (Paris: Vuibert & Nony, 1910), 130–31.

21. Smith, *From Monopoly to Competition*, 8–60; Christina Sonderegger, "Aluminiumzeit," *Kunst+Architektur in der Schweiz* 48, no. 3 (1997): 20–29.

22. Kahren Jones Hellerstedt et al., *Clayton: The Pittsburgh Home of Henry Clay Frick, Art and Furnishings* (Pittsburgh: Helen Clay Frick Foundation, 1988), 54; Mary Brignano, *The Frick Art & Historical Center: The Art and Life of a Pittsburgh Family* (Pittsburgh: Frick Art & Historical Center, 1993), 22, 25.

23. Smith, *From Monopoly to Competition*, 34.

24. "The Present Output of Aluminium," *American Architect and Building News* 36, no. 858 (June 4, 1892): 155.

25. "The Comb Committee," *The Aluminum World* 8 (February 1902): 97.

26. "Incidents in the History of Aluminium," *Aluminium and the Non-Ferrous Review* (January 1936): 142.

27. "L'Aluminium dans l'Outillage du Bureau," *Revue de l'Aluminium* 24 (1928): 623; Gregory W. Smith, "Alcoa's Aluminum Furniture: New Applications for a Modern Material, 1924–1934," *Pittsburgh History* 78, no. 2 (Summer 1995): 52–64.

28. "Chairs of Aluminum on a Quantity Production Basis in a Buffalo Factory," *Furniture Manufacturer* 37 (May 1929): 60–64, 89.

29. *Distinctive Chairs of Aluminum* (Buffalo, N.Y.: Alcraft Aluminum, n.d.), 26.

30. Louise Bonney, "New Metal Furniture for Modern Schemes," *House & Garden* 57, no. 4 (April 1930): 142.

31. Correspondence from Caroline Mathieu, chief curator, Musée d'Orsay, Paris, November 13, 1997.

32. "Aluminum—The War Metal," *House Furnishing Review* 49, no. 4 (October 1918): 42.

33. Douglass B. Hobbs, "Aluminum—A Decorative Metal," *Good Furniture and Decoration* 35 (August 1930): 89.

34. *The Art of the Motorcycle* (New York: Guggenheim Museum, 1998), 174–77.

35. For a history of Jaguar Cars, see www.jaguarcars.com.

36. "4000 Pounds Lighter—Still Greater Capacity," *Aluminum News-Letter* (July 1935): 5.

37. "Metals in the Normandie," *Mining and Metallurgy*, quoted in *Aluminum News-Letter* (September 1935): 1.

38. Quoted in Bonney, "New Metal Furniture for Modern Schemes," 82.

39. John Gloag, "Wood or Metal?" *The Studio* 97 (January 1929): 50.

40. Charlotte Perriand, "Wood or Metal?" *The Studio* 100 (April 1929): 278.

41. "L'Aluminium dans la décoration d'un magasin," *Revue de l'Aluminium* 19 (1927): 458–59.

42. *Revue de l'Aluminium* 40 (1930), supp., n.p.; *Arts décoratifs du XX siècle* (Monaco: Sotheby's Monaco S.A., April 5, 1987), 100, 104.

43. "Les Expositions de duralumin au Salon d'Automne," *Revue de l'Aluminium* 40 (1930): 1294.

44. Marijke Küper and Ida van Zijl, *Gerrit Th. Rietveld* (Utrecht: Centraal Museum, 1992), 209.

45. Brenda Richardson, *Scott Burton* (Baltimore: Baltimore Museum of Art, 1986), 42.

46. "A Partial List of Users of Warren McArthur Anodic Aluminum Furniture," from a Warren McArthur Corporation catalogue (New York, n.d.).

47. David A. Hanks with Jennifer Toher, *Donald Deskey: Decorative Designs and Interiors* (New York: Dutton, 1987), 21, 36, 42, 108; Deskey artist file in Drawings and Print Department, Cooper-Hewitt National Design Museum, New York.

48. Hanks, *Donald Deskey*, 150; "High Fashion in Furniture," *Aluminum News-Letter* (September 1959): 9.

49. Jonathan Lipman, *Frank Lloyd Wright and the Johnson Wax Buildings* (New York: Rizzoli, 1986), 85–91; correspondence from Warren McArthur to Frank Lloyd Wright, The Getty Research Institute for the History of Art and Humanities, Los Angeles, Frank Lloyd Wright Correspondence, M071 A06, M079 B10.

50. "A Problem on Your Doorstep," *Aluminum News-Letter* (July 1935): 6.

51. "Twenty Pounds of Aluminum," *Aluminum News-Letter* (June 1941): 2.

52. Smith, *From Monopoly to Competition*, 191–249.

53. "Aluminum Reborn," *Fortune* 33, no. 5 (May 1946): 108.

54. "Imagineering," *Aluminum News-Letter* (January 1942): 3.

55. "Aluminum—A Basic Material of Tomorrow," *Interiors* 103, no. 11 (June 1944): 68–70, 82.

56. "Housing: The Featherweights," *Time* (May 28, 1945): 89.

57. Joseph Downs, "Design for the Machine," *Pennsylvania Museum Bulletin* 27, no. 147 (March 1932): 115; "Adventure in Aluminum: Russel Wright was one of the first designers to explore its potential," *Industrial Design* 7, no. 5 (May 1960): 54–55; list of contributors and object receipts for the exhibition from the Philadelphia Museum of Art Archives.

58. Alcoa Archive, Box 106, folder 2, Senator John Heinz History Center, Pittsburgh; Philip Johnson, *Machine Art* (New York: Museum of Modern Art, 1934).

59. Dennis P. Doordan, "Promoting Aluminum: Designers and the American Aluminum Industry," *Design History: An Anthology*, ed. Dennis P. Doordan (Cambridge, Mass.: MIT Press, 1995), 158–64.

60. "Your Trip Through the Kaiser Aluminum Show," 1955, Walt Disney Archives, Burbank, California.

61. Joseph Giovannini, "Seat of Authority," *I.D. Magazine* (May 1999): 44–47.

62. Deyan Sudjic, *Ron Arad* (London: Laurence King Publishing, 1999), 141–45.

63. *Aluminium and Electrolysis* (April 1898): 29.

64. *Aluminum News-Letter* (May 1939): 1; "Aluminum Tablecloths, Handbags and Woven Apparel—What Next," *House Furnishing Review* 99, no. 5 (October 1943): 35; Ruth Carson, "Good as Gold," *Collier's* 122, no. 36 (August 30, 1947): 34.

65. Based on conversations with Jane Adam and British Industrial Graphics.

Figure 68.
**Stuart Basseches,** designer
(American, b. 1960)
**Judith Hudson,** designer
(American, b. 1959)
Biproduct, manufacturer
(American, 1998–present)
I-beam table, 1999
anodized aluminum, powder-coated aluminum, and acrylic
cat. 5.44

Figures 69–70.
**Werner Schmidt** (Swiss, b. 1953)
Alu-Falttisch, fully opened (left)
and partly folded (right), 1987
aluminum
cat. 5.41

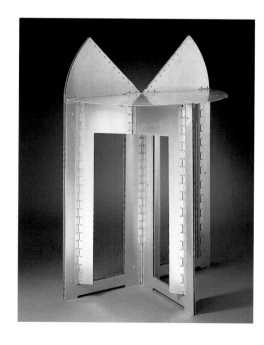

THE LATE GENERAL GERSHOM MOTT.
PHOTOGRAPHED BY J. E. SMITH.—[SEE PAGE 841.]

COLONEL THOMAS L. CASEY, GOVERNMENT ENGINEER OF
THE WASHINGTON MONUMENT.—PHOTOGRAPHED BY BELL.

THE LATE REUBEN R. SPRINGER.
PHOTOGRAPHED BY LANDY.—[SEE PAGE 845.]

B. R. Green, Civil Engineer.　Capt. G. W. Davis, Engineer.　P. H. McLaughlin, Superintendent.　　　Col. T. L. Casey, Government Engineer.　James Hogan, Rigger.　Lewis O'Brien (colored).

572 FEET HIGH—SETTING THE CAP-STONE ON THE WASHINGTON MONUMENT.—FROM A SKETCH ON THE SPOT BY S. H. NEALY.—[SEE PAGE 844.]

# A New Metal! Aluminum in Its 19th-Century Context

Robert Friedel

On a wet, chill day in early December 1884, Col. Thomas Lincoln Casey, U.S. Army Corps of Engineers, stood with several companions on a platform almost 555 feet above a spot near the marshy banks of the Potomac River. With only a modicum of ceremony, he directed the setting of a cast metal pyramid atop the tallest human-made structure in the world. The monument to George Washington that the U.S. Congress had begun planning more than a century before was finally completed, and the final piece was this casting weighing almost exactly 100 ounces and standing a mere 8.8 inches high. As remarkable an achievement as the completion of the monument itself was, the making of that modest little cap was worthy of special note, for it was made of aluminum—a metal unknown to most Americans, and as rare and precious a material as most of them would ever see (fig. 71).

Aluminum—the material of soda cans, frozen-food trays, baseball bats, and spacecraft—would hardly seem to a 21st-century observer a likely candidate for precious metal and the suitable finishing touch to America's most famous monument. This, after all, is the most common metal on earth, at least as measured in the makeup of the earth's crust (aluminum is 7.57 percent, compared with, say, iron's 4.7 percent). Its utility in today's world is everywhere, and its low price makes a thoughtless disposability too common as well. But aluminum is, in fact, the newest of the great metals of commerce, a product of 19th-century chemical legerdemain. For most of that century, it was an uncommon and expensive material, made by a handful of persistent experimenters and entrepreneurs (none of whom would ever get rich on aluminum), and more likely to be encountered in a jeweler's window or a science exhibition than an everyday setting. It was a subject of wonder and admiration to more than one generation of beholders, and as such its place atop the great monument was particularly fitting.

The monument's tip was the proud handiwork of the only maker of aluminum in America, William Frishmuth of Philadelphia. Thirty years earlier, while in his twenties, Frishmuth had left his native Germany and set up a shop devoted to the production of specialty materials, especially rare metals. He was particularly well qualified for this uncommon occupation, having studied chemistry in Germany, where he probably spent a year in the laboratory of Friedrich Wöhler, an eminent chemist of his generation and the first man to behold aluminum. Frishmuth

Figure 71.
Installing the aluminum apex on the Washington Monument, December 6, 1884
*Harper's Weekly* (December 20, 1884): 839

reported making aluminum as early as 1859, but for more than a decade he was diverted by service in the Civil War and then by political opportunities that followed. By the mid-1870s, however, he resumed his chemical work and began to experiment with electricity. He established his expertise as an analytical chemist and an electroplater, but aluminum was clearly his foremost interest. By the time Frishmuth was contacted by the engineer officers in charge of completing the Washington Monument, he had three patents on aluminum extraction, and he advertised himself as "the only manufacturer of aluminium and its alloys in the United States and Canada."[1]

When the Corps initially contacted Frishmuth in October 1884 for help with the monument's tip, its officers were well aware of aluminum's qualities, having earlier considered its use in weather-stripping, but they had rejected it as too expensive. This is probably why they initially asked Frishmuth for a casting of bronze or brass that would then be plated with platinum. A metal tip was important, for lightning was an obvious danger to the monument, and thus its tip was to act as an unobtrusive lightning rod, to be grounded through the monument's interior. It was also important to the builders that the metal be resistant to corrosion and not stain the monument's white marble; hence the request for a platinum coating. Frishmuth was contacted more for his expertise in plating than for his work with aluminum, but he was quick to reply to the engineers' query with the proposal for a casting of solid aluminum. After the usual back-and-forth of a government purchase, chief engineer Casey contracted with Frishmuth for the proposed casting at a cost of $75. The Philadelphia chemist was aware that he was attempting a notoriously difficult feat; just a few years earlier, Clemens Winckler had written in *Scientific American* that aluminum was not much used in part because "no one knows how to cast it."[2]

To ensure that Frishmuth made the tip precisely the right size and shape, Casey sent him a wooden model. From this the chemist had an iron mold made. (The customary and easier molding method is to use a sand impression of the object to be formed, but sand adhered to the molten aluminum and thus was not suitable.) After almost two weeks of effort, he was able to telegraph the engineer in Washington that "after hard work & disappointment I have just cast a perfect pyramide of pure aluminium." It was a point of special pride to Frishmuth that he had made the metal from South Carolina ores, thus avoiding the use of any foreign materials for this most American of monuments. He admitted to Casey that the cost was a bit more than he had anticipated, but he hinted at the splendor of the result by reporting that he wanted some time to exhibit the pyramid to the public before sending it on to Washington. He appears to have first sent it to New York. The *New York Times* of November 25, 1884, described Frishmuth's accomplishment at some length and directed readers to Tiffany's, where they could see the casting on display amidst the jewels and objets d'art of that famous establishment. The tip was back in Philadelphia a few days later, for the *Press* of that city announced on November 28, "Aluminium Exhibited: The Apex of the Washington Monument Made of the Strange Metal." The next day a messenger delivered it to Casey in

Washington, who prepared it for final placement by having the sides engraved with the names of the monument's commissioners and engineers, as well as political leaders of the day.[3]

Far above the Washington Monument's admiring crowds, the aluminum apex still sits, largely out of sight, and a bit worn by more than a century of weathering and no small number of lightning strikes. Indeed, soon after its installation, additional lightning protection was installed in the form of large copper spikes girdling the aluminum apex, in large part obviating the novel casting's original purpose. At least twice in the interval since its installation, replicas have been cast and displayed. The modest Philadelphia foundry in which it was made still stands, but its maker is largely forgotten, his pioneering place in the aluminum industry known only to a few metallurgical enthusiasts and admirers.[4] This is largely due to the fact that Frishmuth's triumph was perhaps more an ending than a beginning, marking the close of a kind of "heroic age" for the new metal. Less than two years after Frishmuth's casting, two young chemists, one working in France and the other in a small town in Ohio, discovered a vastly different and superior way of making aluminum and opened the way for the metal to become, not a noble one fit for ceremony and display, but instead a ubiquitous feature of 20th-century life. The real roots of aluminum's history, however, lie in that early heroic experience, which is comprehensible only in light of the metal's nature and the culture that gave it birth.

Figure 72.
Friedrich Wöhler
(German, 1800–1882)
Test tube with aluminum particles, 1845
aluminum and glass
cat. 2.1

Metallic aluminum does not exist in nature. From the mid-18th century, numerous chemists attempted to isolate the suspected metallic base of the major constituent of common clay, alumina. Finally, the Danish chemist Hans Christian Oersted succeeded in preparing what was probably a mixture of potassium and metallic aluminum in 1825. Two years later in Germany, Friedrich Wöhler produced a better, though still impure, product by passing the vapor of aluminum chloride over heated potassium. The result was a powder made of infusible flecks of aluminum as well as a great deal of solid chloride. In 1845, no significant advance having been made in

the meantime, Wöhler improved his results and obtained pinhead-sized globules of the metal. These were large enough to allow him for the first time to determine several basic physical properties of aluminum, particularly its startlingly light weight (fig. 72). There was, however, little response to Wöhler's announcement, certainly no suggestion that his findings were potentially of practical significance.

In 1853, Henri Sainte-Claire Deville (fig. 73), of the Ecole Normale in Paris, while attempting to make a variation on alumina in experiments with aluminum chloride, produced instead sizable pieces of fairly pure aluminum metal. He quickly saw that he had surpassed Wöhler's results and excitedly began exploring further the properties and possibilities of the metal. He was able to determine that it possessed a silvery brilliance, was ductile, could be melted and molded without oxidation, was resistant to at least some acids, and, as Wöhler had pointed out, was marvelously light. In February 1854, he reported his results to the Académie des Sciences, concluding with the announcement that he intended to develop his process on a scale large enough for aluminum to be introduced to the world of commerce.[5]

Deville's ambitions were encouraged by both the Académie and Emperor Napoleon III, who viewed himself as something of a savant as well as a supporter of French enterprise. Over the next few years, Deville worked to improve his method of extracting metal and to lower the cost of materials. His greatest success came

Figure 73.
**Gabriel Jules Thomas,** designer
(French, 1824–1905)
F. Barbedienne Foundry, maker
(French, 1839–1955)
Bust of Henri Sainte-Claire Deville,
designed 1882, cast by 1900
cast aluminum
cat. 2.3

with the substitution of sodium for the potassium both he and Wöhler had relied upon. It was much easier to devise cheaper ways of producing sodium, and so by 1858, Deville was finally able to offer aluminum for sale at Fr. 300 per kilogram, or $27 a pound. Several different companies were started in France in the 1850s, but only that built by Deville and his backers at Salindres in 1860 proved viable. In spite of technical success, however, it quickly became clear that aluminum would not sell itself. A new metal might be a marvel, but it was not going to become a commodity without great effort in promotion and education.[6]

The first book on aluminum appeared quickly, a small work of 1858 entitled *Recherches sur l'aluminium* written by the brothers Tissier, who had assisted Deville. This book, and the obvious problems of selling even the modest output of his plant, spurred Deville himself to prepare a comprehensive description of his processes and the promise of the new metal. His *De l'aluminium* appeared in 1859 and was an extended expression of his faith in the importance of his discoveries. It also reflected the quickly learned lesson that the place of aluminum in the 19th-century material world was far from clear, and would probably be won after significant experiment and failure. "Nothing is more difficult than to admit into the customs of life and introduce into the habits of men a new material, however great may be its utility," Deville admitted.[7]

*De l'aluminium* listed the applications that had been found for the metal: jewelry, lamp reflectors, telescope mountings, a sextant, opera glasses, a clock pendulum, balances and weights, teaspoons and table settings, medals (figs. 74–78), and—one of the first objects fashioned from aluminum—a baby rattle for the Prince Imperial (Highlight 1). Deville was candid about the limitations of this list: "The objects of luxury and ornamentation are varied enough in their form and nature, but all those which relate to the necessities of life and serve everyday needs are, to the contrary, modified only with great slowness." The key hurdle to aluminum's acceptance, he acknowledged, was aluminum's high price: "Employed in art objects and jewelry, aluminum has encountered no serious opposition to its entry into the industrial world, and bit by bit, by dint of handling and use, will naturally take the place which it should occupy, and which besides, it is necessary to note, depends largely on its price of manufacture and sale."[8]

Deville tied his faith in the future of aluminum most closely to two properties of the metal: its lightness and its benign inertness (*innocuité*). The relative lightness of the metal (its specific gravity is one-quarter that of silver, a third that of iron) was understandably marvelous, and is still the source of much of aluminum's perceived technical value. The equal concern with the metal's safety, its resistance to the production of harmful compounds upon exposure to the environment or in ordinary use, is however quite striking. Nineteenth-century cookery apparently posed greater hazards than we might imagine. "At the moment where it is possible," Deville declares, "to replace the copper in cookware by aluminum, one can notice the immediate disappearance of all the accidents which result so frequently from the dissolving of tin or especially copper in our food." The perceived relationship between aluminum and safe cooking did, in fact, become a continuing theme in the metal's

Figure 74.
**Arthur Martin,** designer
(French)
Paris Mint, manufacturer
(French, 1795–present)
Medallion given at birth,
recto and verso, before 1860
aluminum
cat. 2.13

Figure 75.
**Charles Christofle,** designer
(French, 1805–1863)
Charles Christofle et Cie,
manufacturer (French, 1830–present)
Stand, 1858
chased and gilded aluminum
cat. 2.10

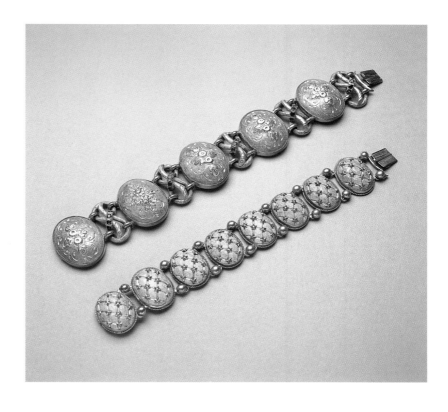

Figure 76.
French
Brooches, ca. 1865–70
aluminum, vermeil, silver,
gilded metal, and carved shell
cat. 2.22, 2.23, and 2.24

Figure 77.
(top) **Victor Chapron**
(French, active 1860–67)
Bracelet, ca. 1865
embossed and chased aluminum,
gilded metal, and garnets
cat. 2.25
(bottom) Attributed to
**Frédéric Milisch**
(French, active 1864–72)
Bracelet, ca. 1865
embossed and chased
aluminum and vermeil
cat. 2.26

Figure 78.
**Armand Dufet**
(French, active 1849–70)
Bracelet, ca. 1860
chased aluminum and gold
cat. 2.27

history, but with much greater complexity and contention than Deville could have anticipated.[9]

The properties of the new metal, however, were not the real source of its future promise to Deville and many of his contemporaries. To expect this to be so is to envision a technological world that is purely practical and utilitarian, but one of the virtues of aluminum's story is the clarity with which it reveals much deeper and more complex motivations in the pursuit of novelty and invention. For mid-19th-century observers, much of the significance of aluminum lay in its unique combining of the properties of noble metals—inertness and safety—with ease of working, general utility, and widely available ores that characterized the common metals. It was not the best or most perfect in either category, but occupied an intermediate ground, hitherto empty. At the very outset of *De l'aluminium*, Deville remarks on the significance of this for him. The very few metals used for "the ordinary needs of life" divided naturally into the precious and the common metals, but "at the moment where chance led me to discover some of the curious properties of aluminum, my first thought was that I had in my hand that intermediate metal" that hitherto had been unknown.[10]

The "intermediate metal" did not mean to Deville the perfect metal, but rather a material that filled a conspicuous void in nature's endowments to humanity, a metal that was common and yet not base, noble and yet not precious. This sensibility was shaped by a worldview that seems rather quaint today, but was at the heart of the 19th-century perspective on technology and its possibilities. This world, on the eve of the publication of Charles Darwin's *Origin of Species* (1859), was a world designed by a benevolent deity, who had strewn about the planet built for His worshippers all the proper elements of a well-provisioned home, only challenging the inhabitants to find and make proper use of His gifts. Aluminum was easily identified in this schema as a necessary missing piece of that divine provisioning. This view also helps to explain the widespread faith that the difficulties in extracting the metal, the primary reason for its high cost, would most certainly in time be overcome by the ingenuity of science and engineering. Associated with this faith was the immediate identification of aluminum as the "silver from clay," as it was dubbed in Deville's display of a few early pieces at the Paris Universal Exposition in 1855. There his samples were placed, tellingly, not with the products of metallurgical industry, but instead among the works of porcelain from the great factory at Sèvres, additional demonstrations of the human capacity to turn common clay into products of art and value.

The efforts of Deville and his compatriots gave aluminum its introduction to notoriety. In September 1855, *Newton's London Journal* reported to its readers: "The recent proposition for the adoption into the arts of the metal aluminium, has afforded to the lovers of the marvellous a new subject for speculation. Already its adaptability to a hundred uses is proclaimed, although the difficulty of obtaining it is at present a bar to its use."[11] The next year, a London lecturer spoke of the use of the metal by surgeons and dentists, and in the making of piano wires and the lining of brass horns, as well as for cooking and tableware. The uses of the future were,

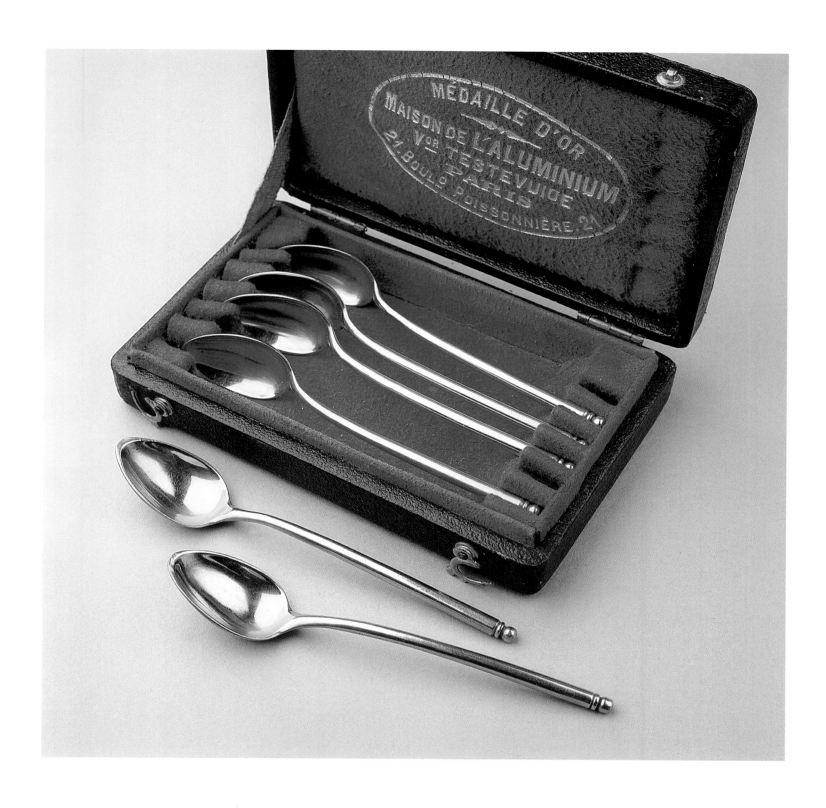

Figure 79.
Société Paul Morin et Cie, later
renamed Société Anonyme
de l'Aluminium, manufacturer
(French, 1857–85)
Maison de l'Aluminium, retailer
(French, 1875–91)
Six spoons in original box, ca. 1880
aluminum bronze, wood, leather,
and gold leaf
cat. 2.29

however, more substantial, for it "may fairly be hoped" that aluminum would soon
be at a price that would compete with iron (the current price was given as £3 per
ounce), and could then be used for many things, such as the roofs of houses, which
were not yet possible. An American lecturer made similar remarks, adding "philo-
sophical apparatus" (scientific instruments), table- and kitchenware, and even the
works of clocks and watches, as likely applications. For such observers, aluminum
was to be a useful substitute for copper, bronze, and other alloys, and occasionally for
silver. They also remarked on the usefulness of aluminum bronze, in which one part
of aluminum was added to about nine parts of copper to produce a hard, workable
alloy with a handsome golden color. Aluminum bronze was easier to place and
appreciate than the pure metal, since it was simply a "better bronze," rather than
a material whose novelty was both an attraction and a problem (fig. 79).[12]

The actual uses of aluminum in these first years were indeed limited, although experimentation was clearly enthusiastic and wide-ranging, within the limits set by the metal's high price. The first public display was at the Paris Universal Exposition of 1855. In the Palais de l'Industrie, among fine objects like Sèvres porcelain, lay several bars of aluminum and silver, allowing the public to compare the two metals and even lift them to comprehend the astounding difference in their weights. In addition, from the famous workshops of Christofle came a small round dish, the spoon and fork of a cutlery set, and a small coffee spoon, all of aluminum. The virtues of aluminum's light weight were displayed by an anemometer with aluminum blades. An even more striking scientific instrument, however, was an aluminum balance arm made by the Collot Frères, France's premier makers of scientific balances. This was entirely of aluminum, with the exception of small parts such as screws and counterweights. Accompanying the balance was a set of precision weights that could be used for measuring precious metals or in chemical assays. The balance arm itself survives, one of the treasures of the Musée des Arts et Métiers in Paris (figs. 80–81). It is perhaps the oldest extant aluminum object, with the exception of a medal bearing a likeness of Napoleon III (now in Munich's Deutsches Museum; Highlight 1) made by Deville a few months earlier and sent to Friedrich Wöhler as a token of respect.[13]

In spite of aluminum's placement in the Paris Exposition among objects of art and fine craftsmanship, the more important identification of the metal in these first years was as a product of science, yet another treasure wrested from nature's storehouse. In an unsigned article in the *National Magazine* in mid-1857, this perception comes through particularly clearly: "A new metal! yes, a new metal, and yet as old as the world we inhabit!… there seems now good reason to expect that, at no very distant day, it will come into general use, for all purposes to which it may be found adapted." The foundation for this confidence lay largely in the ubiquity of aluminum's ores, "found in all the varieties of clay… even in old bricks and broken crockery!" But behind this practical consideration lay a deeper faith:

> By the aid of science man is gradually extending his dominion over material nature, and subjecting the elements more and more to his service; and within the last few years some of his proudest achievements have been accomplished, as in the art of *photography* and the *electric telegraph*. And just now, in this new metal, so long concealed in every hill-side, and even in the very dust of our streets, science seems about to make over to the arts one of her occasional bestowments, by which both the knowledge and power of our race are, at an instant, so widely increased.[14]

To men and women of the mid-19th century, aluminum's real importance lay in its place as part of the onward march of science and progress. Its eventual importance in the scheme of the human world was assured less by obvious applications or the fulfilling of apparent needs than by aluminum's neat fit into an ideology of advancement and of nature put more and more at the service of humanity.

Figure 80.
Collot Frères, later renamed
Collot-Longue et Cie
(French, 1848–1920)
Balance arm of scale, 1855
aluminum
cat. 2.2

Figure 81.
Collot Frères, later renamed
Collot-Longue et Cie
(French, 1848–1920)
Scale with 1855 balance arm,
before 1920
aluminum, brass, and agate
cat. 2.2

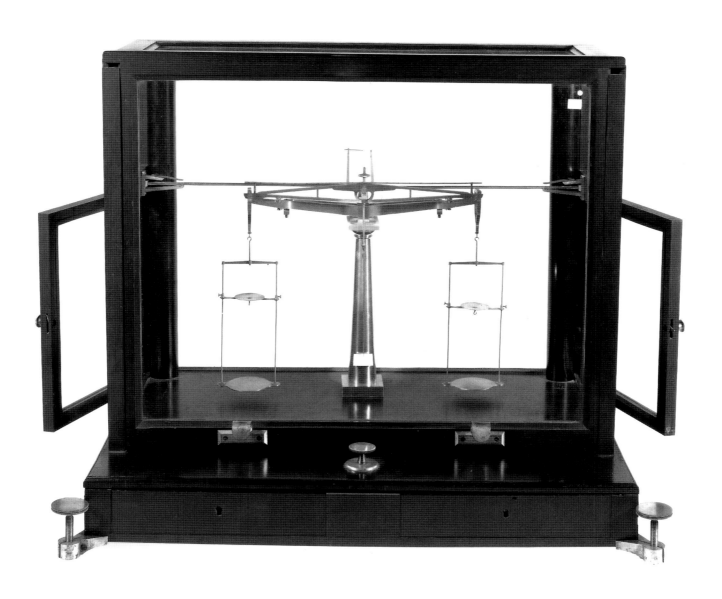

To a degree, the interest voiced in aluminum during the 1850s was the product of a fad, a spontaneous reaction to the announcement out of France of yet another marvel of science. The relative dearth of similar pronouncements over the next couple of decades shows how shallow the excitement really was for most observers. The technical community as a whole was not much engaged by the new metal during the 1860s and '70s; not a single American patent for aluminum extraction was issued between 1858 and 1879. This is not to say that the metal was forgotten. In this period it found itself a small but comfortable niche within the materials of *objets de fantaisie*, in Deville's words. A popular British book on mining and metallurgy of 1866, for example, included a chapter on aluminum, speaking of "the beautiful silvery-white metal…now being made and sold in jewellers' shops, and used for ornamental purposes as brooches, bracelets, chains, &c. &c."[15]

If we seek an example of these *objets de fantaisie* from this period, we could hardly ask for one more curious than the extraordinary fan in the political history collection of the Smithsonian Institution in Washington, D.C. (figs. 82–83). The fan itself is an impressive object, made of ornate aluminum blades and folds of painted silk. The blades are cast with openwork, filigree, and chasing, each ending with an eight-pointed star, the top and bottom blades having elongated points. No wonder one reference to the fan was as an *éventail défensif*, for such a device could have been quite lethal in the hands of an outraged holder. The metal features eagles, stars, and violets in rampant profusion, but their impression is mute in comparison to that made by the striking decorations of the silk and attached paper. Glued to each side of the nine interior blades is a small oval photograph of a woman—the

Figures 82–83.
**August Edouard Achille Luce,** designer
Crespo De Borbon, retailer (Cuban)
Fan, recto and verso, ca. 1866
aluminum, silk, and paper
cat. 2.19

reigning queens of Europe—along with a smaller oval identifying her country.
A number of these are now missing, but the effect is still remarkable. On the silk itself
is a striking set of images, quite unconnected with royalty and, indeed, in ironic
contrast to those photographs: on one side, seven small painted scenes, two depicting
famous battles of the American Civil War, and five chronicling the assassination of
Abraham Lincoln at Ford's Theatre in Washington on the evening of April 14, 1865,
the ensuing pursuit of John Wilkes Booth, and his eventual death in a burning Virginia
barn almost two weeks later. On the other side, Napoleon III's empress, Eugénie,
is given pride of place in the photographic portraits, and above her is a portrait
of Lincoln flanked by figures, flowers, and a banner with the words, "To the martyr
of his country Abraham Lincoln" (fig. 83). Flanking this are depictions of monumental columns, American eagles, and, on paper, the words and music of a Spanish
song in Lincoln's honor. It is only around the song sheets that one can find small
clues to the fan's origins: "Fan protected with privileges (patented)," "Sold by Crespo
De Borbon," "The patent plated fan," "Habana-Paris."[16]

As incredible as it seems, this fan, probably made in Cuba sometime between
1865 and 1867, is not unique, but a registered product of its Havana manufacturers.
The Smithsonian's donor claimed that the fan was made for the Paris Universal Exposition of 1867, but no record of the fan can be found in the exhibition's catalogues,
even though fine fans were by no means neglected in the records of that fair (one
guide declared that the section devoted to the display of this Parisian specialty was
"certainly one of the most curious and most charming" of the exhibition). Another
copy of the fan is in the Oldham Collection at the Museum of Fine Arts, Boston,

although it is not identical; many of the portrait medallions, for example, are of Civil War generals rather than European royalty and, significantly, the fan's blades are of brass rather than aluminum. Perhaps the Smithsonian's fan was indeed a special, precious version made for the Paris fair, but no contemporary documentation supports that claim. It is, nonetheless, remarkable testimony to aluminum's exotic allure in its first decade of availability.[17]

Following a pattern sustained through the second half of the 19th century, another international exposition was held in Paris eleven years after that of 1867. By 1878, aluminum's novelty had passed, even though its applications were little changed. A German observer wrote:

> The Paris Exhibition offered us, in 1878, the view of the maturity of the aluminum trade. We have passed out of the epoch in which aluminum was worked up in single specimens, showing the future capabilities of the metal, and see it accepted as a current manufacture, having a regular supply and demand, and being in some regards commercially complete.

Despite this upbeat note, most observers in the 1870s made note of aluminum's use in opera glasses, candlesticks, and mustard spoons, and noted its promise in more extensive applications if its price were lowered (figs. 84–88). Estimates of world aluminum production during this period are little more than 5,000 pounds per year, all of it French, and never at less than $12 a pound. In retrospect, reference to "the maturity of the aluminum trade" would seem to have been awfully premature.[18]

This sense of complacency about aluminum was largely lacking in America, where the early 1880s marked a resurgence of interest in new ways of making and using the metal. This same period, to be sure, saw an astonishing expansion of American interest in all things mineral and metallurgical, and to a degree aluminum —a very small part indeed of a much larger picture—was the beneficiary of this. The opening of the West and the exploitation of its vast resources had, by this time, become the centerpiece of the American story, and the prospect of wealth and opportunity from each bit of metal had become the common property of every ambitious man and woman. The empires of steel in the East, represented foremost by Andrew Carnegie's great mills, and the bonanzas of silver, copper, tin, lead, and even a bit of gold in the West made manifest the metallurgical promise of the future. The U.S. government recognized this with the sponsorship of great surveys in the West and the formation in 1879 of the U.S. Geological Survey.

The Survey began issuing an annual report, *Mineral Resources of the United States*, in the early 1880s, and it is here that we can see the emergence of aluminum in the American consciousness. The first report, for 1882, treated the metal in a one-page entry that reviewed the developments of the 1850s, with the conclusion that "the industry, if it may be called so, has remained substantially as Deville's experiments left it." Even this sober assessment was not free of the usual references to a different future, one in which "aluminum can be produced so cheaply as to make practicable its speedy application to the purposes for which its properties

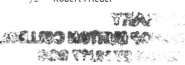

Figure 84.
French
Opera glasses, ca. 1865–75
aluminum, glass, nickel-plated
metal, and mother-of-pearl
cat. 2.17 and 2.18

Figure 85.
Probably French
Cruet stand, ca. 1875
aluminum, metal,
and cut glass
cat. 2.30

Figure 86.
Probably French
Stand, ca. 1875
cast and chased aluminum
cat. 2.11

Figure 87.
Morizot, printer (French, active second half of 19th century)
Prayer book, ca. 1870
chased aluminum, aluminum leaf, board, paper, electroplated silver, and velvet
cat. 2.16

peculiarly adapt it." Tellingly, this edition of *Mineral Resources* gave greater attention to metals like iridium, tellurium, and cobalt than to aluminum.[19]

As the decade of the eighties went on, however, the situation changed. Interest in aluminum quickened, and the government survey was anxious to report progress. The combined report for 1883–84 spoke of increased use of the metal and the growing familiarity of manufacturers and the general public with its properties. William Frishmuth's manufacture in Philadelphia came in for special mention, where his reported output was given as 1,800 ounces at 50 cents per ounce. The next year, Frishmuth's output almost doubled, according to the Geological Survey report, and the use of aluminum for small articles increased, although there was no progress on price. Indeed, the usual avowal—heard by this time for thirty years in both Europe and America—was echoed, that "its use would be greatly extended when its price was lowered." Greater attention was drawn to the emergence of an aluminum-alloy industry in America, largely relying on processes that could not produce the pure metal, but could make aluminum bronze or alternative alloys with some economy. The important figures in this connection were the Cowles Brothers, whose Electric Smelting and Aluminum Company harnessed newly reliable electric dynamos to make electrochemical methods of extraction practicable for the first time.[20]

In the historiography of aluminum, the year 1886 stands out as the turning point, the year in which the young Charles Martin Hall, working in the family toolshed in Oberlin, Ohio, discovered how to use electricity to make pure aluminum metal, while his contemporary Paul Héroult made a similar discovery in France. But this sense of breakthrough and momentous change existed only in the hopes and ambitions of the young men and their backers, not in the world at large. The Geological Survey reported that in 1886 the Cowles Brothers manufactured some 50,000 pounds of aluminum bronze and several thousand pounds more of ferro-aluminum alloy, but the pure metal was proving, according to the government report, as elusive as ever. "Little, if any, aluminum of domestic production was in the market," *Mineral Resources* reported for 1886, although imports of several hundred pounds were counted, consistent with the pattern of the decade. In hindsight, the irony is heavy in the conclusion that despite reports in the press of "proposed new methods of extracting aluminum,…no material improvement in that respect was made public in 1886."[21]

The skepticism of government observers becomes a bit more understandable in light of the series of scattered press reports published in the first half of the 1880s promising the coming of cheap aluminum. Under the heading "Current Foreign Topics," the *New York Times* for December 14, 1882, carried a notice from London announcing excitement over a new process for the production of aluminum. The following day, the paper sought out the opinion of a New York metals assayer: "Mr. Platt said that so-called cheap processes for making this metal had been discovered periodically in the last 15 years, but he had never seen any of them amount to anything." This skepticism notwithstanding, the *Times* reported a month later on British advances by reprinting an item from the *Pall Mall Gazette* proclaiming that within a few months' time aluminum "will be manufactured into articles differing as widely as a tablespoon and a ship's anchor." In 1884, spurred by the display at Tiffany's of William Frishmuth's apex for the Washington Monument, the newspaper reported on the first native contribution to the field. The report was headed "Rival to Older Metals," and went on to speak of "aluminium" in large quantities at low prices. "The philosopher's stone has not been sought," the *Times* reporter waxed, "with greater eagerness than this cheap process." A few months later, in April 1885, the *Times* reported an English process by "a man named Webster" (this was the same process that had been the subject of the 1882 article) that similarly promised large-scale production of the metal: "When the secret of its cheap production is generally known a revolution in the metal world will be the certain result." In June of 1886, the paper reprinted sizable portions of an article from the journal *Power*, which spoke of "How and When Iron's Great Rival May be Made Available." And the next month, the newspaper reported the announcement out of Newport, Kentucky, of a process that would make aluminum as cheap as copper. Again the praises were effusive: "The importance of this discovery can be judged when it is recollected that aluminum is the most generally diffused metal on earth, and has all the beauty of silver, besides being non-tarnishing, non-corrosive, more lasting than silver, with only one-fourth its weight."[22]

The article that the *Times* excerpted from *Power* is worth a moment's further notice, for it distilled many of the central themes that aluminum represented for its enthusiasts throughout the first will-o'-the-wisp phase of the metal's technological history. The promise of a rival for iron is one often heard in aluminum's heralding, a promise that resonated strongly in a Victorian age of iron horses, iron ships, and iron men. The properties of aluminum were popularly held up as vastly superior to iron's: capable of being hardened "till the diamond is its only rival," capable of being made into wire so fine or sheet so thin "that the gold-beater alone can do the work," of great strength and durability, highly conductive, capable of taking a polish to rival silver. Aluminum was, to some observers, the ultimate metal, not just Deville's intermediate metal filling an available space in the rank of useful materials, but a metal to rival all others. The cheapness of iron alone allowed it to hold the field, but the ultimate promise of cheap aluminum seemed tantalizingly near, when it would take the field as the "new iron."

But how could aluminum really be a "new iron"? In truth, the technologies of the 19th century—the railroad, the steam engine, the machine shop and furnace and textile mill—were not technologies that would find aluminum very useful, even when its price, within just a few years of the remarks of the newspaper reporters, the display in Tiffany's window, and the careful skepticism of the government observers, fell to within the range long promised. Indeed, in retrospect, the frequent declaration of aluminum's universal application may have been a tacit recognition of the fact that the unique properties of the metal had no specific utility, beyond those narrow ones already exploited, and could command no sure market. For thirty years prior to the establishment of the modern aluminum industry at the end of the 19th century, the output of a small French factory satisfied fully the Western world's real demand for the metal. The absence of a sure and ready market for cheaper aluminum became quickly apparent to the inventors and investors who set out to make the metal's promise for the future into reality.

The story of Charles Martin Hall's discovery of the long-sought process for making "cheap" aluminum has been told often, and is, in fact, one of the classic tales of inventive American youth using brains and persistence to "strike it rich" in the fields of technology. It is worth summarizing here, however, for the three decades of aluminum's earlier history—from Deville's laboratory to the apex of the Washington Monument—now provide a useful context for both questions and conclusions about the real meaning of Hall's achievement. It is appropriate, also, to focus on Hall at the expense of the other discoverer of the electrolytic reduction of aluminum, Paul T. L. Héroult, for the effort to make pure aluminum into a metal of everyday life and commerce drove Hall much more clearly than it did Héroult, who was more easily diverted down the road of alloys, in the fashion already demonstrated in America by the Cowles Brothers.[23]

Both Hall and Héroult were only twenty-two years old when they made their separate discoveries, and both were men from comfortable backgrounds with good educations. It is Hall, however, who comes across most clearly as a young, untested man—his boyish looks, in fact, never left him. He was the eager student, the one who enthusiastically listened as his Oberlin College chemistry professor, Frank F. Jewett, extolled the virtues and promise of aluminum. Jewett had himself studied in Wöhler's Göttingen laboratory years earlier, and so knew the metal of which he

spoke—and all the promises made in its name. Even before he entered college, Hall had become an enthusiastic home experimenter, setting up a laboratory first in his mother's kitchen, which was then banished to the woodshed next to the house. Jewett encouraged a range of experiments, although apparently the quest for a cheaper aluminum caught his pupil's eye quite early. Another source of encouragement was Hall's sister Julia, who possessed, like her brother, considerable self-education in chemistry. Julia was a source of more than simply sisterly encouragement, for she also assisted in keeping records and, mindful of the requirements for valid patenting, was a careful observer of key experiments. Whether her assistance went beyond this, as some have claimed, is really impossible to say from surviving records, for Julia Hall shared a traditional Victorian notion of the woman's expected role in worldly endeavors and always deferred to her brother's reputation.[24]

Hall's efforts began in the traditional place, with Deville's chemical processes and efforts to make them easier and cheaper, either by using less costly materials or by making them easier to scale up. He had no more success than earlier experimenters in this direction, but he was as observant as he was persistent, and thus, particularly after his graduation from Oberlin in June 1885 and subsequent dedication to a full-time attack on the problem, he knew as much about aluminum and its compounds as anyone. The circumstances of the mid-1880s gave Hall an important new avenue that earlier experimenters had considered but not found fruitful: the prospect of economical electricity. Deville himself, among others, had been aware of the theoretical possibility of extracting aluminum from its ores by electric current, but when this proved difficult, he, and others after him, abandoned this route due to the high expense of electric current, which, until the 1880s, was available for all practical purposes only from batteries. The introduction of electric lighting at the beginning of the decade, however, was accompanied by the development of reliable and cost-effective electrical generators, and this changed the economic equation significantly. Hall himself still had to rely on expensive batteries (his process consumed about a pound of zinc electrodes for each ounce of aluminum extracted), which he prepared in his workshop, but he was aware that outside of the laboratory such a costly expediency would not be necessary.

Some metals can be produced by passing an electric current through a water-based solution of one of the metal's salts—thus a solution of copper sulfate in water is electrolyzed to produce metallic copper, the basis of copper plating. For many metals, however, electrolyzing aqueous solutions will not yield metallic products, but simply other compounds of the metal. This is true for aluminum compounds, as Hall discovered when he passed a current through a solution of aluminum fluoride in water, producing not metal but aluminum hydroxide. The well-known path around this problem—useful for producing sodium from common salt, for example —was to melt the metallic compound and electrolyze it directly, rather than in solution. That this could be done for aluminum had been demonstrated in the 1850s, but a great deal of fuel is required to melt most aluminum salts, and so there was no promised economy from this. Perhaps the easiest compound to melt is a mineral known as cryolite, a so-called double fluoride, being compounded with both

Figure 88.
F. Barbedienne Foundry
(French, 1839–1955)
*Venus de Milo*, 1889
cast aluminum
cat. 2.64

aluminum and sodium. Cryolite is not a common mineral, nor is it particularly easy to prepare, so the well-known fact that molten cryolite could be electrolyzed to produce metallic aluminum was of limited value. The key to Hall's discovery lay in determining that alumina, the oxide of aluminum, can be dissolved in molten cryolite, and from this solution can be electrolyzed to produce the metal.

In experiments conducted February 9–10, 1886, Hall determined that alumina did indeed dissolve easily in molten cryolite, and less than a week later he proceeded to attempt electrolysis from this solution. The initial result was not promising, for at the electrode where he hoped to find shiny metal, he discovered only a brittle grayish coating. A bit of investigation convinced Hall that this was in fact silicon, which could only have come from the clay crucible in which he had carried out the experiment. He then prepared a crucible from carbon graphite, melted the cryolite again (adding a little aluminum fluoride to lower the melting point still further), dissolved some alumina, and ran the current through for several hours. The result

Figure 89.
**Louis Blum** (American, b. 1909)
Box holding Charles Martin Hall's globules of aluminum, ca. 1935
aluminum
cat. 2.34

consisted of some shiny globules of pure metal, which Hall quickly determined (and asked Jewett to confirm) were pure aluminum (figs. 89–90).

Hall's discovery was not a complex one, but it required a particularly intimate knowledge of the materials he was using and of how their properties related to one another. It was also, it is important to note, driven by a specific faith that pure aluminum metal, not some alloy of it, was the only serious goal worth pursuing. The depth and significance of this faith became clear in subsequent months, as Hall attempted by several routes to find the necessary backing to transform his laboratory discovery into a useful industrial process. For some time he negotiated with the Cowles Brothers, successful makers of aluminum bronze, to have them take over the Hall process and make the pure metal their key product. But the brothers, while clearly worried about what Hall's work might mean for their own business, did not share his confidence that pure aluminum had greater marketability than the alloy.[25] Finally, in mid-1888, backers in Pittsburgh created the Pittsburgh Reduction

Company and, in a pilot plant on Smallman Street, Hall successfully converted the laboratory process into an industrial one, using new Westinghouse dynamos to produce the needed electricity.

By mid-1889, Hall's plant was producing fifty pounds of aluminum bars a day, and the product stacked up in the Smallman Street shop for lack of buyers. Perhaps due to some nervousness over their patent situation, the Pittsburgh manufacturers were not quick to publicize their work. One of the earliest notices appeared in the *New York Times* for September 3, 1889, in which the manager of Hall's plant was quoted as admitting they were using "a patent process, in which electricity is a great factor,… but we have managed so far to keep the matter out of the newspapers." He continued to emphasize that they did "not wish to furnish any further information," but mentioned that their metal cost about $4 per pound, and had found a number of uses, most specifically taking "the place of silver leaf in sign painting." That summer, Hall had written to a company salesman out on the road, "I hope you will succeed in getting a good outlet for our metal. It is piling up in the office at the rate of thirty to fifty pounds a day," and went on to praise the sample match cases he had seen. With the exception of a steadily developing, though small, demand from local steelmakers, who had discovered that a small amount of aluminum added to molten steel vastly improved castings by reducing gas formation, the fact is that the Pittsburgh Reduction Company simply had no markets to speak of.[26]

This is the final ironic truth of aluminum's 19th-century "prehistory." All the celebration and all the promise for the future were in terms of a belief in the wisdom of Providence, in the necessity of nature's storehouse being made complete with a metal both common and noble, in a magical combination of properties that seemed to foretell a new era in the arts. But the celebration and the promise were quite independent of technological or economic reality. Indeed, one of the virtues of aluminum's story is its eloquent denial of the role of necessity in the expansion of our material world. Aluminum indeed was one of the most useful and appreciated materials of the 20th century, and in many ways promises quite as grand, if different from the 19th-century ones, were fulfilled in skyscrapers, airplanes, spacecraft, and a host of other applications, most not even dreams to the Victorians.

## Notes

1. George J. Binczewski, "The Point of the Monument: A History of the Aluminum Cap of the Washington Monument," *JOM* 47, no. 11 (1995): 20–25; "Washington Monument Correspondence, 1884–1935," ALCOA Records 1888–1990, Historical Society of Western Pennsylvania, Pittsburgh.

2. Binczewski, "The Point of the Monument." Geo. W. Davis to Phosphor Bronze Smelting Co., Philadelphia., August 29, 1884, W. Frishmuth to Th. L. Casey, October 25, 1884, and October 31, 1884, in "Washington Monument Correspondence, 1884–1935," ALCOA Records; Clemens Winckler, "Aluminum," *Scientific American Supplement*, no. 192 (September 6, 1879): 3058.

3. Frishmuth to Casey, October 31, 1884, and November 12, 1884 (telegram copy), in "Washington Monument Correspondence, 1884–1935," ALCOA Records; Binczewski, "The Point of the Monument."

4. See George J. Binczewski, "A Monumental Point: World's First Big Aluminum Casting," in *Light Metals 1993* (Warrendale, Pa.: Minerals, Metals & Materials Society, 1993), 723–30.

5. R. E. Oesper and P. Lemay, "Henri Sainte-Claire Deville, 1818–1881," *Chymia* 3 (1950): 210; Joseph W. Richards, *Aluminium: Its History, Occurrence, Properties, Metallurgy, and Applications*, 3d ed. (Philadelphia: H. C. Baird & Co., 1896), 9.

6. Richards, *Aluminium*, 11–14; Claude J. Gignoux, *Histoire d'une entreprise française* (Paris: Hachette, 1955), 28–30; "Aluminium," *Scientific American* 13, no. 34 (May 1, 1853): 265.

7. Henri Sainte-Claire Deville, *De l'aluminium: Ses propriétés, sa fabrication et ses applications* (Paris: Mallet-Bachelier, 1859), 140 (author's translation).

8. Ibid., 140.

9. Ibid., 147.

10. Ibid., 1.

11. "Aluminium, the So-called New Metal," *Newton's London Journal of the Arts and Sciences* (September 1855), reprinted in *Journal of the Franklin Institute* 61 (1856): 27.

12. J. Barlow, "On Aluminium," *Proceedings of the Royal Institution* 2 (March 14, 1856): 221–22; W. J. Taylor, "Aluminium: The Progress in Its Manufacture," *Proceedings of the Academy of Natural Sciences of Philadelphia* (January 27, 1857): 15–16; "The New Metal, Aluminum," *National Magazine* 10, no. 29 (May 1857): 448.

13. Jean Plateau and Elise Picard, "Le fléau de la balance ou les apparitions de l'aluminium: L'Exposition de 1855," *Cahiers d'Histoire de l'Aluminium* 23 (Winter 1998): 9–28.

14. "The New Metal, Aluminum," 448.

15. John H. Pepper, *The Playbook of Metals* (London: G. Routledge and Sons, 1866), 395.

16. Division of Political History, National Museum of American History, Smithsonian Institution, catalogue no. 323481, catalogue and accession records.

17. Louis Enault, "Les éventails," *L'Exposition universelle de 1867 illustrée* (Paris, 1868), vol. 2: 179; see also Eugene Rimmel, *Recollections of the Paris Exhibition of 1867* (Philadelphia: J.J. Lippincott & Co., 1867), 71; *Les merveilles de l'Exposition universelle de 1867* (Paris: Imprimerie Générale de Ch. Lahure, 1867), 207–08; Anna C. Bennett, *Unfolding Beauty: The Art of the Fan* (Boston: Museum of Fine Arts, 1988), 256. My thanks to the Smithsonian's Shelley Foote and the MFA's Elizabeth Ann Coleman for assistance in locating the Oldham Collection fan.

18. Winckler, "Aluminum," 3058. See also *The Technical Educator: An Encyclopedia of Technical Education* (New York: Cassell, Petter and Galpin, 1872), 55; Robert Routledge, *Discoveries and Inventions of the Nineteenth Century*, new ed., rev. (London: G. Routledge and Sons, 1877), 509–10. Data from "Aluminum Applications," in ALCOA files, no. 8-296.

19. U.S. Department of the Interior, U.S. Geological Survey, *Mineral Resources of the United States, 1882* (Washington, D.C.: U.S. Government Printing Office, 1883), 445.

20. *Mineral Resources of the United States, 1883–1884* (Washington, D.C.: U.S. Government Printing Office, 1885), 658–59; *Mineral Resources of the United States, 1885* (Washington, D.C.: U.S. Government Printing Office, 1886), 390–91.

21. *Mineral Resources of the United States, 1886* (Washington, D.C.: U.S. Government Printing Office, 1887), 220–21.

22. *New York Times*, December 14, 1882, 1; December 15, 1882, 2; January 15, 1883, 2; November 25, 1884, 3; April 14, 1885, 5; June 20, 1886, 6; July 8, 1886, 3.

23. Héroult's European patents, very similar to Hall's American ones, did become the basis for the aluminum industry in Europe. Héroult himself, while vigorously promoting his process and the aluminum industry generally, had the same central role in Europe that Hall did in America, due in part to the diversity of corporate interests that defined the European industry (different firms were formed to exploit his patents in Switzerland, France, Germany, and Britain within a few years of one another), and in part to Héroult's own broader interests in electrochemistry generally, which soon expanded beyond aluminum itself. See Gignoux, *Histoire d'une entreprise française*, esp. 65–90.

24. Perhaps the best recent account of Hall's work is Norman C. Craig, "Charles Martin Hall—The Young Man, His Mentor, and His Metal," *Journal of Chemical Education* 63 (1986): 557–59; the fullest treatment is the somewhat fawning book by Junius Edwards, *The Immortal Woodshed* (New York: Dodd, Mead, 1955).

25. The Cowles Brothers did begin manufacturing pure aluminum in 1891, but were found in infringement of the Hall patents in 1893. Undaunted, they challenged the Hall–Pittsburgh Reduction Company monopoly by referring to an obscure patent (issued to one C. S. Bradley) they had acquired that covered parts of the reduction process at a very broad level. They won their suit in 1903, but instead of continuing to compete with the Pittsburgh company, they chose instead to take penalty and royalty payments and leave the manufacturing to the Hall interests. The basic patents expired in 1909, but the Pittsburgh company managed to maintain its near monopoly status in much of the American aluminum industry for more than thirty years longer. See Donald H. Wallace, *Market Control in the Aluminum Industry* (Cambridge, Mass.: Harvard University Press, 1937), 5–6; Charles C. Carr, *ALCOA: An American Enterprise* (New York: Rinehart & Co., 1952), 217–32.

26. "Aluminum Making in Pittsburg," *New York Times*, September 3, 1889, 2; Edwards, *Immortal Woodshed*, 110.

# From Precious to Pervasive: Aluminum and Architecture

Dennis P. Doordan

The phrase "from precious to pervasive" identifies one of the central themes of this exhibition: aluminum's transformation from its origins as an exotic metal to its present status as one of the ubiquitous materials of modern life. How aluminum became so pervasive in the realm of architecture is the subject of this essay and, if it reads as a success story, the reader should be forewarned: the story is not a simple one. When, for example, a piece of cast aluminum was set into place on top of the Washington Monument in December 1884, it marked a singular application of a still novel material for a unique monument in the nation's capital. Today, it is a rare new construction that fails to employ aluminum in some capacity, ranging from structural frames, exterior sheathing, window sashes, and door frames to gutters, grilles, ductwork, and diffusers. Indeed, for promotional purposes, buildings like Minoru Yamasaki's Reynolds Metals Regional Sales Office in Southfield, Michigan, completed in 1959, intentionally were conceived as veritable catalogues of aluminum applications (figs. 91–92).

While it is easy to demonstrate aluminum's pervasiveness in the built environment, explaining how this came to be presents a peculiar challenge for architectural historians. A catalogue of representative examples of architectural uses of aluminum quickly grows unwieldy and, in the absence of any interpretive framework, is surprisingly uninformative. Historical reviews of particular building types, monographic treatments of individual architects, and discussions of the design ideologies of various architectural movements each shed some light on the story of aluminum's emergence as a major factor in the material culture of modern design. Yet none of these approaches pursued in isolation adequately conveys the whole story. No single building, architect's work, or design manifesto can be identified as the seminal contribution that propelled aluminum to its preeminence among modern architectural metals. Instead, this essay argues that aluminum became a pervasive material because of the momentum generated by the disparate forces that, considered together, inform the conceptualization and fabrication of the built environment. These "forces" include an aluminum industry eager to identify new markets, manufacturers of building components searching for ways to improve their products (and hence profits), and the design theories that inspired modern architects.

Figures 91–92.
**Minoru Yamasaki** (American, 1912–1986)
Exterior and interior views of Reynolds Metals Company Regional Sales Office Building, Southfield, Michigan, 1959

## Early Applications

The aluminum tip installed on the Washington Monument measured barely 9 inches high and 5.5 inches wide at its base. Weighing 100 ounces, it was the largest piece of aluminum yet cast in the United States. Still a rare metal in the early 1880s, guidebooks included descriptions of the new metal's properties and origins, and the piece was displayed prominently in the window of Tiffany's in New York City prior to being set in place atop the stone monument in Washington, D.C. Yet the monument's engineers maintained that their decision to use aluminum was based on the inherent properties of the metal and not its novelty value. As one of the earliest histories of the Washington Monument noted:

> The aluminum point which caps the monument is not only ornamental but
> useful.... Aluminum was selected because of its lightness, one third that of
> copper, and its freedom from oxidation. As long as the obelisk remains the
> point will be as bright as it is at present.[1]

As a light-weight metal that resists oxidation, aluminum emerged as a substitute for more traditional and heavier metals on both sides of the Atlantic. For the church of San Gioacchino in Rome, started in the early 1890s but not finished until 1898, the architect Raffaele Ingami employed aluminum to sheathe the exterior of the dome (fig. 93). The choice of this novel material had little impact on the architectural style of the church. Ingami's design conforms to the historicist *stile cinquecentesco* then popular in Italian architecture.[2] Ingami's choice of aluminum suggests another aspect of the early history of architectural aluminum that should be noted. San Gioacchino was erected to celebrate the jubilee anniversary of Pope Leo XIII's ordination to the priesthood and, although aluminum was a truly modern material in the sense that it was a product of 19th-century scientific and technological advancements, in the 1890s it had yet to acquire an association with polemical concepts of modernity that would subvert the traditional symbolic program of a church.

Figure 93.
**Raffaele Ingami** (Italian, 1836–1908)
San Gioacchino, Rome, early
1890s–1898

Figures 94–95.
**Burnham and Root** (American,
1873–91)
Elevator enclosures (top) and
staircase (bottom) from Monadnock
Building, Chicago, 1889–92
*The Winslow Brothers Co. Ornamental
Metals* (catalogue, Chicago, 1893),
plates 60 and 124

As a light-weight metal, aluminum proved to be a viable substitute for iron, bronze, and brass in ornamental metalwork and building fixtures. Only a few years after the completion of the Washington Monument, the Winslow Brothers Company of Chicago was providing cast aluminum sections for use as staircases, elevator doors, ventilating grilles, and other fixtures in commercial buildings. Burnham and Root's Monadnock Building, opened in 1892 in Chicago, most often cited by historians as marking the end of a tradition of load-bearing masonry construction for tall commercial buildings, was one of the first to feature extensive applications of cast aluminum for interior metalwork (figs. 94–95). Along with the Monadnock, Burnham and Root specified aluminum for the Venetian and the Isabella buildings, completed in 1891 and 1893 respectively.[3] In 1898, *The Aluminum World*, an industry trade publication intended to disseminate information concerning the growing number of uses for aluminum, reported Daniel Burnham's assessment of the new metal.

> I like the metal because it does not rust, but looks well and when used awhile it turns a dull color and does not shine like nickel-plated brass or other metals. It also lasts where you put it. There will be a future for aluminum in replacing wood in certain buildings. However, that will be largely a question of price and conditions. I believe aluminum is a useful metal for architects at present and that it will be more useful in the future.[4]

Despite an appreciation of its inherent properties and a genuine interest in potential applications, price, as Burnham indicated, remained the biggest obstacle facing aluminum producers in the late 19th century. Only when aluminum could be produced at a price competitive with more established building materials could it truly begin to have a significant impact in contemporary design practice. Solving the problem of achieving production on a large scale at a competitive price was left to the metallurgists and engineers working in the aluminum industry's research centers and production facilities.

## Modern Architecture and Design Theory

American architects were not the only ones beginning to incorporate aluminum in new buildings. European architects were gaining exposure to the novel metal as the nascent European aluminum

industry began to produce and market its product. A particularly noteworthy chapter in the history of aluminum was being written in Vienna during the first decade of the 20th century. In 1906, a new Postal Savings Bank was inaugurated in the Austrian capital (fig. 96). Designed by Otto Wagner, the building quickly gained international recognition as a bold step beyond the eclectic historicism that had characterized civic architecture in Vienna during the last half of the 19th century. Wagner employed aluminum throughout the building in a variety of ways. The thin marble panels of the building's exterior were anchored into place and the anchoring bolts capped with raised aluminum heads. The bolt heads served double duty as both a constructional device and a decorative motif. In the interior, aluminum was used as a cladding material in the major public space, and various fixtures and grilles were fabricated of aluminum rather than the more conventional iron or wood. Several years earlier, in 1902, Wagner had used aluminum extensively in his design of a facade for the news agency Die Zeit (fig. 97). Here the combination of aluminum panels, prominent electric lighting fixtures, and stylish lettering signaled a bold departure from the traditional architectural decorum of Vienna.

Wagner's designs initiated an important new chapter in the story of aluminum. Unlike the tentative first uses described above, they represent far more than the

Figure 96.
**Otto Wagner** (Austrian, 1841–1918)
Exterior of Austrian Postal Savings Bank, Vienna, 1904–06

Figure 97.
**Otto Wagner** (Austrian, 1841–1918)
Full-scale replica of 1902 Die Zeit facade, 1984–85
aluminum and glass
cat. 2.72

piecemeal substitution of one material for another on the basis of commercial considerations of price, weight, or availability. Once inside the Postal Savings Bank, one has the sense that a cultural threshold has been crossed. Wagner's work projected an aura of modernity that clearly distinguished his efforts from the traditional designs characteristic of previous generations. In Wagner's designs, aluminum becomes emblematic of a new, self-consciously modern approach to building. Wagner outlined this new approach in his important work *Modern Architecture*, first published in 1896. Two years earlier, he had been appointed professor of architecture at the Vienna Academy of Fine Art. Because of his prestigious position at the academy, *Modern Architecture* was treated as an important statement by architects and critics across Europe. In *Modern Architecture*, Wagner assailed 19th-century conceptions of architectural practice predicated on the eclectic use of historical styles and traditional construction as inadequate for contemporary needs. His architectural treatise was inspired, Wagner wrote, by one idea: "that the basis of today's predominant views on architecture must be shifted, and we must become fully aware that the sole departure point for our artistic work can only be modern life."[5]

The feeling among progressive European architects that they needed to "shift" their thinking about architecture in light of the evolving conditions of modern life manifested itself with a growing sense of urgency during the first decades of the 20th century. Young architects responded enthusiastically to Wagner's admonition that

> All modern creations must correspond to the new materials and demands of the present if they are to suit modern man; they must illustrate our own better, democratic, self-confident, ideal nature and take into account man's colossal technical and scientific achievements, as well as his thoroughly practical tendency —that is surely self-evident.[6]

Iron, steel, and reinforced concrete tended to head the list of materials identified as modern by architectural theorists of the era. Although aluminum was seldom referred to directly, it is important to note the broad outlines of the emergent modern design ideology because it establishes the context within which the equation of a material—aluminum—with a cultural phenomenon—modernity—could be developed.

Early efforts to define modern architecture revolved around three broad themes. The first dealt with the social agenda of the new architecture. Architects sought to harness the productive capacity of modern industry in order to solve social problems caused by inadequate housing and services in new neighborhoods springing up in the expanding peripheries of urban centers. No longer would the design community be satisfied with its traditional role as provider of services to society's wealthy elite. The second theme was focused on the formal problem of the new architecture. Informed by the artistic experimentation of avant-garde painters and sculptors, architects explored bold new conceptions of what constituted beauty in the built environment. Historically sanctioned precedents for composition were rejected in the quest for an architecture conceived as the image of a new world. Finally, progressive architects were concerned with tectonic issues and the possibilities inherent in new building technologies. New materials, modes of problem analysis, and methods of construction were reconfiguring the building trades and the design community's understanding of the art of construction.

Throughout the 1920s and '30s, progressive architects discussed these themes and, as a result, long before the architectural community as a whole was familiar with aluminum, a climate receptive to new materials and techniques was being nurtured. Writing in 1924, for example, the German modernist Ludwig Mies van der Rohe neatly described the commitment of his generation to the radical reconfiguration of design practice.

> It is not so much a question of rationalizing existing working methods as of fundamentally remolding the whole building trade. So long as we use essentially the same materials, the character of building will not change...Industrialization of the building trade is a question of material. Hence the demand for a new building material is the first prerequisite. Our technology must and will succeed in inventing a building material that can be manufactured technologically and utilized industrially... It will have to be a light material whose utilization does not merely permit but actually invites industrialization.[7]

Although Mies van der Rohe did not mention aluminum specifically, passages such as this are important in understanding the way modern architects were attempting to link design issues with tectonic concerns regarding materials and construction. Much of aluminum's eventual popularity as a building material can be attributed to the perception on the part of progressive architects that it was an extremely viable answer to the "question of material" posed by Wagner, Mies van der Rohe, and other European modernists.

Of course, avant-garde European architects were not the only ones interested in the relationship between new materials and new ways of thinking about architecture. Frank Lloyd Wright, an inspirational model for architects in both Europe and America, continued his tireless campaign to reorient the direction of American architecture. The appropriate handling of building materials was an important theme in Wright's architectural thinking and, although he seldom discussed aluminum specifically, he promoted a climate of engagement with materials that would prove decisive in the story of aluminum. In the late 1920s, *Architectural Record* magazine published a series of articles by Wright entitled "In the Cause of Architecture." Each article addressed a different aspect of Wright's philosophy of modern architecture, including the proper handling of different materials. The October 1928 installment dealt with sheet metal. In the late 1920s, aluminum still occupied a marginal position in the theoretical elaboration of modern architecture, and Wright's essay focused primarily on copper and steel. Wright was clearly attuned to evolving conditions in the market for building materials; he treated lead and wrought iron as craft materials traditionally worked by hand and speculated about the impact of modern machine production on metal fabrication. "The machine is at its best,"

Wright wrote, "when rolling, cutting, stamping or folding whatever may be fed into it."[8] What could most easily "be fed into the machine" according to Wright was metal in the form of sheets that could then be crimped, folded, or stamped "as an ingenious child might his sheets of paper."[9] While Wright hailed standardization and industrial fabrication of metal components as the wave of the future, he also assailed what he considered the dull and lifeless result of buildings conceived strictly in terms of optimizing technological processes and commercial interests. For Wright, it was up to the architect to breathe life into inert matter and imbue his buildings with grace and beauty.

Wright discussed aluminum only in passing, and the skyscraper design he included as an illustration in his essay was clad in copper, not aluminum. However, his belief that exploiting the tectonic possibilities of modern metals meant understanding the properties of sheet metal fashioned by stamping and rolling presses is important. Wright's essay is also revealing of the state of knowledge among progressive architects in the first third of the century. A survey of the defining literature for early-20th-century modernism—the manifestoes, programs, and treatises of the new architecture—reveals an enthusiastic climate of engagement with new materials and processes but a notable lack of understanding concerning specific production technologies. The possibility of extruding thin sections of a light-weight metal like aluminum rather than stamping it in the form of metal sheets, for example, attracted little attention. Indeed, one of the critical chapters in the story of aluminum occurred outside the realm of avant-garde architectural movements and modernist design programs.

### Kawneer and the Metal Storefront

Otto Wagner, Ludwig Mies van der Rohe, and Frank Lloyd Wright belonged to the professional world of architecture and each, in his own way, illuminates a different aspect of this international culture of high design. Francis Plym, although a graduate of the University of Illinois School of Architecture, belonged to a different world: the world of business. In 1906, Plym was awarded a patent for his invention of a resilient metal window molding that could be used to frame large shop windows. It was the first of more than one hundred patents Plym would be awarded before his death in 1940. The story of Plym's first patent conforms to a formula common in stories of invention: a flash of insight followed by a period of research and development. In Plym's case, the "eureka" moment occurred one evening in 1904 as he and his wife were waiting for a streetcar in Kansas City. Plym noticed that the wooden molding surrounding a large shop window was already showing signs of rot even though the building was still under construction. Moisture condensing on the shop window was running down the plate-glass front and soaking the wooden molding. The heavy plate glass exacerbated the problem: the sodden wooden molding was bulging under its weight. What if, Plym wondered, wooden molding could be replaced with a stronger material less susceptible to rot and the

Figure 98.
**Francis Plym** (American, 1869–1940) Storefront of F. Johnson Co., with Plym's patented window frames, Holdrege, Nebraska, 1905

Figure 99.
**Francis Plym** (American, 1869–1940) Drawings of F. Johnson Co. Storefront, 1905
Thomas Stritch, *The Kawneer Story* (Niles, Mich.: Kawneer Company, 1956), 66

Figure 100.
Extruded elements for window assembly
John Peter and Paul Weidlinger, *Aluminum in Modern Architecture*, vol. 2 (Louisville, Ky.: Reynolds Metals Company, 1956), 59

FIRST COMPLETE
KAWNEER FRONT

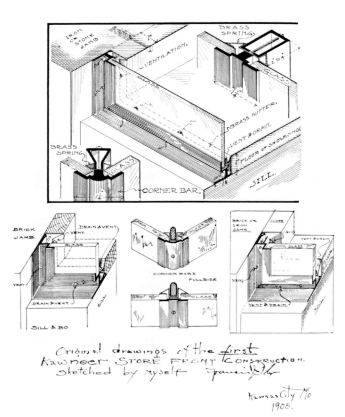

Original drawings of the first
Kawneer STORE FRONT Construction.
sketched by myself  Francis Plym

Kansas City Mo
1905.

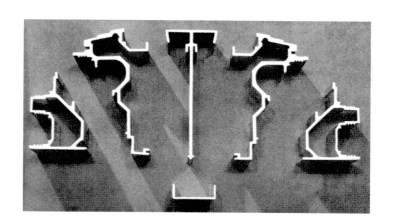

moisture could be drained away from the frame? Two years later, Plym patented his answer to this question: a sturdy, flexible metal frame with an internal gutter to carry away any condensed moisture dripping from the glass and openings in the frame for ventilation to equalize the pressure on both sides of the plate glass.[10]

In 1906, Plym founded the Kawneer Company, which quickly became an industry leader in the development of metal windows and storefronts (figs. 98–99). Kawneer's early product lines were manufactured primarily of copper, bronze, or steel. It was not until the early 1930s that aluminum began to make significant inroads in the fabrication of commercial windows and storefronts. Once it was available in quantities sufficient to meet the demand, aluminum quickly came to dominate Kawneer production, and by 1937, fully 75 percent of Kawneer's product line was aluminum-based. Plym and his engineering staff were oriented to manufacturing commercially viable, pragmatic solutions to specific problems. The ethos of the Kawneer Company was fundamentally different from that of the architectural avant-garde. Plym was a businessman, not a prophet of cultural revolution, and the Kawneer sales staff sold a product line, not a radical avant-garde vision of a bold new world. An early Kawneer catalogue of steel window sashes extolled the virtues of metal windows in terms guaranteed to appeal to a businessman's concern for gauging the return on an investment.

> Money-saving—because Kawneer Solid Steel Windows will reduce maintenance, insurance, and lighting to a minimum cost. Money-making—because the natural lighting and ventilation obtained will increase the efficiency of employees so that every dollar spent will bring in a maximum return.[11]

It was in this climate of pragmatic entrepreneurialism that important technical advances in metal fabrication were worked out. Instead of stamping sheet metal for various types of panel systems, Kawneer began to work with extruded shapes. In the extrusion process, metal is heated and then forced through a die. The extrusion process produces long pieces of uniform cross section that can be fitted together and assembled quickly on site. It is an ideal process for fabricating elements with the type of complex sections required for large commercial window frames (fig. 100). Technical expertise, not design theory, is the essence of the Kawneer contribution to the story of aluminum.

## The 1930s

The 1930s were a decisive decade in the story of aluminum. Kawneer's increasing reliance on aluminum is indicative of a larger shift in the status and perception of aluminum during this decade. Prior to 1930, the emergence of aluminum as an essential metal for modern architecture was a disjointed story of special cases like the Washington Monument or the gradual substitution of aluminum for other metals in subsections of larger buildings like shop windows. As late as 1926, the selection

Figure 101.
**Henry Hornbostel**
(American, 1867–1961)
Spire of Smithfield United Church,
Pittsburgh, 1926
*Revue de l'Aluminium* 20 (1927): 487

of aluminum for the spire of the Smithfield United Church in Pittsburgh can best be understood as a novel demonstration of aluminum's ability to replicate traditional metalwork in a lighter material (fig. 101). Around 1930, however, the disparate elements of the aluminum story began to coalesce, and a new consciousness of the advantages and potential of aluminum took root among architects and engineers. In the United States, articles outlining the advantages of aluminum and reporting on new construction featuring the metal appeared with increasing frequency in architectural journals.[12] No longer treated as a novelty, aluminum—along with the other so-called white metals: chrome-steel, chrome-nickel steel, and Monel (a nickel-copper alloy)—was now being discussed as an integral part of light-weight exterior wall systems designed to replace heavier traditional masonry construction. The appeal of new types of wall construction for tall office buildings and other types of commercial architecture was clear. In December 1931, Harold Vader of Holabird and Root's office succinctly made the case for aluminum in an article published in *Architectural Record*:

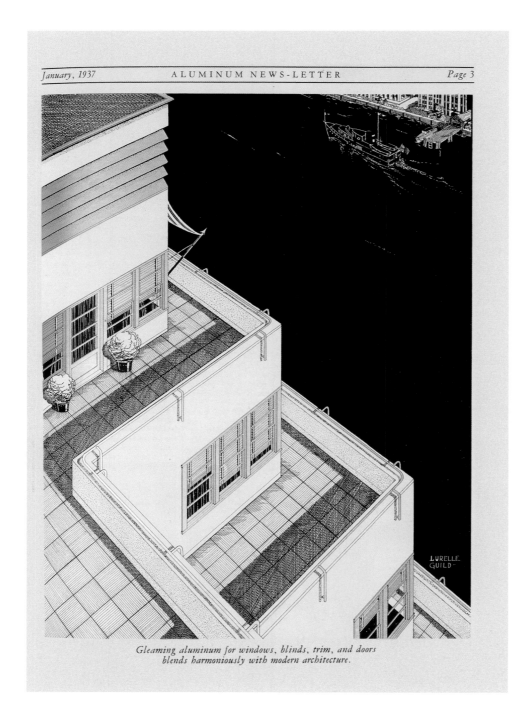

LURELLE GUILD

*Gleaming aluminum for windows, blinds, trim, and doors
blends harmoniously with modern architecture.*

Today, a change in this [i.e., traditional] type of wall construction seems logical
from a purely economic standpoint. The skyscraper came into being in order
to make a profitable use of the land. By the use of metal facades it is possible to
reduce the thickness of the walls 8 to 10 inches. This saving in wall area can be
converted into rentable space. With the load on the structural-steel frame reduced
by the use of metal in place of masonry, a material saving can be effected in
the supporting structure. Again, a metal facade can be erected in approximately
one-fifth the time required to erect a masonry wall. This results in economies
particularly beneficial to the owner. Not only are erection costs reduced but the
owner is able to capitalize on his investment in much shorter time.[13]

"Space, weight, and time" became the mantra invoked consistently in discussions of aluminum. Using aluminum in a variety of ways, an architect could increase
the space available for lease, reduce the weight of the structure, and accelerate the
pace of construction. In the calculating world of commercial architecture, any material

Figure 102.
**Lurelle Guild** (American, 1898–1986)
Drawing published in Alcoa's
newsletter to promote the use
of aluminum
*Aluminum News-Letter*
(January 1937): 3

or construction system that improved financial return on investment proved attractive. But the appeal of space, weight, and time extended beyond the world of business architecture, and aluminum became an important nexus linking manufacturers of building materials with architects committed to the modernist design program.

One of the best period examples of the growing links between designers and industry was the Aluminaire House conceived for the 1931 Allied Arts and Building Products Exhibition in New York City (figs. 103–106). Designed by the partnership of A. Lawrence Kocher and Albert Frey, Aluminaire was a model house constructed at full scale inside the exhibition hall of Manhattan's Grand Central Palace.[14] This prominent venue, and the fact that the 1931 Building Products show was held jointly with the Architectural League's fiftieth anniversary exhibition, ensured a huge audience for the event. A. Lawrence Kocher was the managing editor of *Architectural Record* and the designer, in 1929, of one of the first poured concrete houses on the East Coast, in Fairfield County, Connecticut, for the writer Rex Stout. His partner, Swiss born and educated Albert Frey, had emigrated to the United States a year earlier after working in the atelier of Le Corbusier, one of the leading figures of the European architectural avant-garde.[15] Kocher and Frey conceived of the Aluminaire House as a demonstration of the principles not just of contemporary architecture but of modern living as well. They had written previously of the health benefits of increased lighting levels in living and working environments,[16] and the Aluminaire House featured huge windows, open terraces, and an array of novel lighting fixtures. According to Frey, the name "Aluminaire" suggested modernity: "It had aluminum in it, you know, and it was very airy. And also luminaire means light."[17] The house was constructed with aluminum-pipe columns carrying a steel floor deck and clad with thin aluminum panels fixed to the frame with aluminum screws and washers.

In the design of the Aluminaire House, we see the convergence of various themes in the development of modern architecture. The Aluminaire House reflected the aesthetic preference of the modernist avant-garde, and the design owes an obvious debt to Le Corbusier's Five Points of the New Architecture.[18] But it owes an equally large debt to the various American manufacturing companies that subsidized the cost of the Aluminaire. Alcoa provided the aluminum, and the steel floor decks were manufactured by Truscon. McClintic-Marshall Corporation (a subsidiary of Bethlehem Steel) and Westinghouse also contributed components. While Kocher and Frey were promoting the cause of modern architecture, companies like Alcoa and Truscon were interested in promoting innovative uses for their products (fig. 102). Thus progressive architects and large corporations began to find those areas where their interests intersected. By harnessing the building technology and productive capacity of American industry, architects like Kocher and Frey hoped to create a new market for affordable, hygienic, modern housing. The economies of scale generated by modern systems of mass production could make possible, so the argument went, the manufacture of prefabricated components that could be quickly assembled on-site at less than the cost of traditional construction. Clean, bright, and efficient homes could be brought within the financial reach of thousands of potential home buyers unable to afford traditional designs. The modernization

and rationalization of design practice in light of the properties of new materials and the engineering logic of new construction techniques would herald the arrival of a truly modern architecture. Without, of course, endorsing the more radical aspects of the modernist social agenda, large corporations engaged in the building trades began to see the potential for new applications of their products and services. The dream of good design and good business working hand in hand to build a better world for all was alive and well in the 1930s.[19]

At the conclusion of the Allied Arts and Building Products Exhibition, the architect Wallace K. Harrison purchased the Aluminaire House. Harrison had the model house dismantled and moved to his estate on Long Island, where it was reassembled in an altered and enlarged form. (In 1987, the building was purchased by the New York Institute of Technology for eventual reassembly on the school's Long Island campus.) As one of the few American buildings featured in Philip Johnson and Henry-Russell Hitchcock's epoch-defining International Style exhibition and publication sponsored by the Museum of Modern Art in 1931, the Aluminaire House marks an important milestone in the development of modern architecture in America. It was not, however, the only such attempt to promote both modern architecture and modern building materials. Throughout the 1930s, world fairs, trade shows, publicity pavilions, and magazine competitions kept the ideals of modernism in design before international audiences in the form of demonstration buildings, fair pavilions, and corporate or industrial exhibits.

While Kocher continued his efforts to promote modern architecture in his role as a journalist, Frey pursued a career in professional practice. He eventually settled in Palm Springs, California, where he designed a series of strikingly modern

Figures 103–106.
**A. Lawrence Kocher,** architect (American, 1885–1969)
**Albert Frey,** architect (American, b. Switzerland, 1903–1998)
Aluminaire House, Syosset, New York, 1931

homes adapted to local conditions. His own house there, originally designed in 1940, demonstrates his unfailing commitment to the modern ideals expressed so clearly in the Aluminaire House. Frey enlarged his house twice, first in 1948 and again in 1953. The 1953 remodeling involved the addition of a dramatic second-floor bedroom, round in plan and punctuated by a series of circular windows overlooking the pool and patio (fig. 107). The house combines the easy indoor-outdoor lifestyle typical of this desert resort community with a particularly flamboyant sense of modern design. The round dining-room table and the treads of the stairs leading to the bedroom were suspended from the ceiling by aluminum rods, for example, and the bedroom walls were covered with a yellow tufted vinyl fabric with electric-blue vinyl drapes for the windows.[20] Striking, flamboyant, unconventional: these describe Frey's residential designs beginning with the Aluminaire House.

In order to appreciate the inroads made by aluminum in the design and construction of homes, it is necessary to put Frey's accomplishments in some perspective. Where Frey and his modernist colleagues saw "clean lines" and expressive forms as evocative of an age of jet aircraft and fast cars, many home buyers saw only strange, mechanistic designs closer in spirit to the world of factories than to the cozy domestic world of hearth and home. Dramatic and uncompromising, modern designs were exceptions rather than the rule in postwar American residential architecture despite their prominence in many architectural histories of the period. Commenting on the attitude assumed by most postwar American home buyers to the machinelike imagery of modern architecture, the architectural historian Sidney Robinson noted:

> On a basic level, the average home buyer in the 1940s considered machines to be portable objects made of metal, like those used in the war, and they considered them to be incompatible with houses, which were wooden and immobile—that is, the new materials and forms could be successfully transferred to automobiles, as General Motors stylist Harley Earl knew, but home was a place of refuge and traditional comfort.[21]

Aluminum enjoyed its greatest success in residential architecture when it adapted to, rather than dictated, the image of the house. The true aluminum house of the second half of the 20th century is more than likely a Colonial Revival design with aluminum siding, window sashes, gutters, and downspouts (fig. 108).

## The 1940s, '50s, and '60s

Between the experimental panels of the Aluminaire House and the familiar aluminum siding and trim employed extensively in American suburban developments during the 1950s and '60s stand the watershed years of World War II. No discussion of aluminum is complete without an account of the war's impact. During the 1940s, the aluminum industry in America was completely transformed.[22] The military

demand for this strong, light-weight metal was voracious, and the increase in
aluminum production far outpaced increases in the production of other metals.
In the United States, much of the cost of increasing production was subsidized
by the federal government. Prior to the outbreak of war, Alcoa was the primary
producer of aluminum, and the company stood ready to purchase new production
facilities in order to maintain its monopoly. However, in 1945, the United States
government initiated antitrust actions against Alcoa and announced its intention
to promote the development of competition within the aluminum industry.
R. J. Reynolds, long a consumer of aluminum for cigarette packaging, and the Kaiser
Company, a shipbuilding and heavy-construction concern, purchased production
facilities from the federal government and entered the market as primary producers
of aluminum. As a result, a new competitive climate existed within the industry
as its leadership struggled to find new postwar markets to absorb their increased production capacity. The industry's efforts in the 1950s and '60s can be characterized
in two broad categories: (1) a return to strategies interrupted by the Great Depression
and the war, and (2) the aggressive promotion of a new climate of engagement
with aluminum.

Advances in the design and construction of curtain walls for tall office
buildings illustrate the first type of effort. Curtain walls are a characteristic element
of large-scale modern construction. Rather than being self-supporting, curtain
walls are attached to the structural frame of a building. A promising architectural
application for aluminum in the 1930s and '50s was as spandrel panels in curtain-
wall construction. In multistory buildings, spandrel panels fill the space between
the top of a window in one story and the sill of the window in the story above, thus
masking the exterior edges of floor slabs and the parapets supporting the window

Figure 108.
American Colonial Revival house
with aluminum siding, 1960s
*Aluminum by Alcoa* (Pittsburgh:
Alcoa, 1969), 54

assemblies. Because of its light weight, aluminum proved to be a popular material for use in the curtain walls of tall office buildings. Shreve, Lamb & Harmon, architects of New York City's Empire State Building, used aluminum spandrels in their design. The story of the Empire State Building has been amply documented; conceived and executed in a span of only eighteen months, every aspect of the building's design was rationalized to facilitate rapid erection.[23] The soaring steel frame is sheathed with a curtain wall consisting of window and aluminum spandrel panels installed between limestone slabs. R. H. Shreve described the design of the building's curtain wall in the July 1930 issue of *Architectural Forum*. The windows are supported on a masonry parapet that, in turn, rests on the steel beams of the building's structural frame (fig. 109). Aluminum spandrel panels, measuring 4 feet 6 inches high and 5 feet wide, mask the exterior face of this beam-and-window assembly. In his account of this window-spandrel-wall arrangement, Shreve speculated about alternate configurations and lamented the lack of time available to explore such possibilities.

> In later developments of this construction arrangement it should be possible to support the window from the steel frame of the building and, if desired, to make the metal spandrel integral with the window and its support; it may be possible to later also omit the apron wall (between the window sill and the fireproofing of the structural steel). But for the very rapid construction program of the Empire State Building it has seemed wiser to adopt the tried method of window support as one involving no experiments and no non-standard manufacture—and affording the greatest assurance of rapidity and continuity of execution.[24]

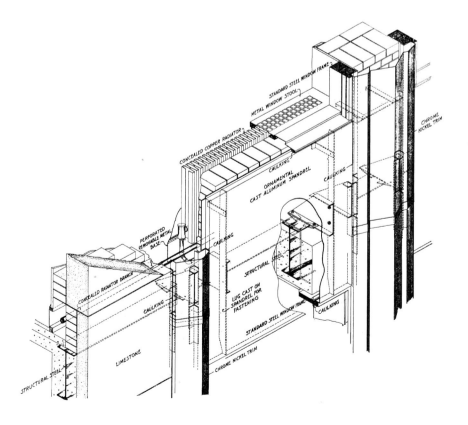

Figure 109.
**Shreve, Lamb & Harmon**
(American, 1929–ca. 1990)
Curtain-wall diagram of the
Empire State Building
*Architectural Forum* (July 1930): 101

Because of the economic depression of the 1930s, and later, the war, the development of an integral window-spandrel system would wait until the renewal of large-scale commercial building in the 1950s.

When building activity did resume after the war, the aluminum industry was ready to demonstrate the viability of new panel systems for curtain-wall construction. In late 1950, Alcoa began construction of its new headquarters in downtown Pittsburgh. Designed by the New York firm of Wallace K. Harrison and Max Abramovitz, the office tower was conceived as a proving ground for products fashioned of aluminum (figs. 110–111). As a result, the architects enjoyed luxuries denied to the Empire State Building's designers: time and the support required to research new design ideas. Alcoa's Research Laboratories in New Kensington, Pennsylvania, conducted scores of tests on various proposed systems and components, and full-scale mock-ups of various curtain-wall designs were erected for testing. Company literature hailed the final building as a "thirty-story demonstration of aluminum's usefulness, economy, and beauty…showing aluminum to be at once practical and economical in almost every phase of building construction."[25] The Alcoa Building curtain wall consists of thin aluminum panels measuring 6 feet by 12 feet containing a 4 foot 2 inch by 4 foot 7 inch opening for the insertion of an aluminum framed window (fig. 113). Rather than resting on a masonry parapet wall, the panels could be bolted directly to the structural steel frame. Aluminum's light weight meant the panels could be quickly hoisted into place and assembled with a minimum of heavy construction equipment (fig. 112). The erection of the Alcoa Building provided incontrovertible evidence that what architects like Mies van der Rohe (fig. 114) had dreamed about in the 1920s and '30s—the industrialization of building—was now an accomplished fact.

Aluminum producers were eager to capitalize on this fact, and Alcoa, Reynolds, and Kaiser—the three primary American producers of aluminum—did everything they could to educate architects about the properties of aluminum, provide them with the necessary technical information, and stimulate a climate of creative engagement with the material. Beginning in the early 1950s, the big three aluminum producers generated a veritable flood of publications, films, recorded interviews with prominent designers, and awards promoting architectural applications for aluminum and celebrating recent designs. One of the most impressive efforts was mounted by the Reynolds Metals Company. In 1956, Reynolds published a lavish two-volume set entitled *Aluminum in Modern Architecture* dedicated to architectural uses of aluminum.[26] Volume 1 featured over one hundred buildings erected since the end of the war. Volume 2 served as an engineering reference book with tables and performance standards to assist architects in specifying aluminum products. Reynolds recorded interviews with prominent architects extolling the virtues of aluminum and released the result as an LP record, which was widely distributed to schools

Figure 110.
**Harrison & Abramovitz**
(American, 1945–76)
Alcoa Building with spire of Smithfield United Church in foreground, Pittsburgh, 1950–53

Figure 111.
**Harrison & Abramovitz**
(American, 1945–76)
Lobby of Alcoa Building, Pittsburgh, 1950–53

Figure 112.
**Harrison & Abramovitz**
(American, 1945–76)
Workers hoist first panel of Alcoa Building, Pittsburgh, 1950–53

Figure 113.
**Harrison & Abramovitz**
(American, 1945–76)
Curtain-wall diagram of Alcoa Building, Pittsburgh, 1950–53
*Aluminum on the Skyline*
(Pittsburgh: Alcoa, 1954), 17

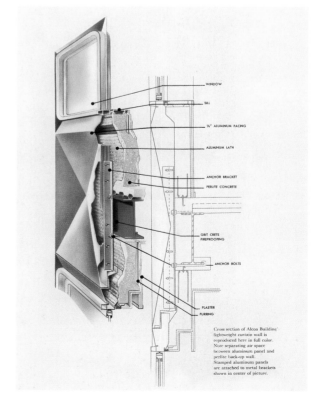

Cross section of Alcoa Building lightweight curtain wall is reproduced here in full color. Note separating air space between aluminum panel and perlite back-up wall. Stamped aluminum panels are attached to metal brackets shown in center of picture.

of architecture as well as practicing architects. That same year, Reynolds insti-
tuted a new international award in the amount of $25,000 to be given annually
to the architect who had made "the most significant contribution to the use
of aluminum, aesthetically or structurally, in the building field."[27] At a dinner held
in New York City to announce the award program, Gaylord Harnwell, president
of the University of Pennsylvania, delivered a keynote address that left no doubt
that the award heralded a new alliance between the design profession and the
aluminum industry.

> A great award such as that which has just been announced as a memorial
> to Richard Samuel Reynolds represents a torch which is handed by industry to
> genius in the Olympics of artistic and technical achievement. The product of
> human labor provides this stimulus to excellence in art...Through this award
> art and labor are wedded to science and technology, which are the charac-
> teristic facets in the culture of our age.[28]

The implications of this new alliance were significant for the design profes-
sions. In his 1930 article describing the Empire State Building's curtain wall, Shreve
had lamented the lack of time and resources needed to investigate alternate
configurations for the window and spandrel assemblies. Twenty-five years later, it
appeared the aluminum industry was dedicated to providing as much assistance
as the architectural community needed. Beyond providing technical support, the
aluminum industry worked hard to create within the architectural community
a professional milieu in which an acceptance of aluminum as the logical material
for a myriad of building applications was so pervasive as to be unquestioned.
In architectural offices across America, thinking about aluminum was becoming
part of thinking about architecture and design.

Alcoa, Reynolds, and Kaiser were hardly alone in their efforts to focus the
attention of the architectural community on specific materials and building systems.
By late 1942, both architects and companies involved in the construction industry

were looking ahead optimistically to the renewal of peacetime building. Rather than being an empty chapter in the history of 20th-century architecture, the war years were filled with proposals, programs, and competitions that sought to focus architects' attention on the possibilities of postwar building.[29] Although Kawneer's wartime efforts were redirected from commercial storefronts to military needs, the company never lost sight of its original market niche or the need to cultivate designers' interest in its products. The October 1942 issue of the professional journal *New Pencil Points* carried an announcement of a Kawneer-sponsored competition for "Store Fronts of Tomorrow." Potential entrants were encouraged to "demonstrate originality and imagination" in the design of a group of five storefronts.[30] Winning entries, along with jury comments, were published in the February 1943 issue of the journal.[31] Of particular importance was the group of professional architects assembled by Kawneer and *New Pencil Points* for the competition. William Lescaze, a leading American modernist architect of the 1930s, served as Kawneer's professional advisor for the competition and set the terms of the design problem. The jury consisted of Roland Wank (Knoxville, Tennessee), Morris Ketchum, Jr. (New York), Samuel Lunden (Los Angeles), Frederick Bigger (Washington, D.C.), and Ludwig Mies van der Rohe (Chicago). All the prize-winning entries wedded a modern design aesthetic of transparent volumes articulated by thin planes and cantilevered roofs to commercial considerations for product display, signage, and special lighting effects. In a brochure issued by Kawneer reporting on the results of the competition, the company described how the competition fit within its program of wartime research and development.

> In the complete break with the past that has come with the war, we have studied the experience of nearly half a century of store front manufacture. We have analyzed what architects have done with store front design over a period of years, checked the current thinking of architects and designers with a national store front competition, welcomed new ideas from many sources. In addition, the intensive wartime production of aluminum assemblies has taught us many new things about the fabrication and finishing of aluminum which will be reflected in our postwar products.[32]

Thus, in the midst of war, Kawneer was preparing for the resumption of peacetime building and doing so in ways that drew influential leaders of the architectural profession into the effort by soliciting their design ideas and publicizing the results.

For its part, the architectural profession was only too happy to work with both the aluminum industry and major product manufacturers. As industrialization in building—eagerly embraced as an idea in 1924 by Mies van der Rohe—became a reality (at least in the field of commercial architecture), it brought with it a new set of concerns for the architectural community. If critical building components and systems were now to be fabricated at the factory and assembled on-site rather than constructed in the traditional manner, then important design decisions were being made in situations that bore little resemblance to traditional patterns of architectural practice. In a building culture increasingly oriented to technologically sophisticated systems of industrially produced elements, critical design parameters were now being established not in the architect's studio but in the research and development laboratories of major producers like Alcoa, Reynolds, and Kaiser or on the production lines of manufacturers like Kawneer. The implications of emerging construction practices were clear—and slightly ominous—for many architects. As the original director of the Bauhaus in Germany and, beginning in 1938, chair of the Department of Architecture at Harvard University, Walter Gropius was one of the most respected postwar leaders of the architectural profession. In *Scope of Total Architecture*, he expressed genuine concern about the marginalization of the design profession in the industrialized world:

> So it might be appropriate to investigate how far our professional framework
> fits the condition of our time…Let's see if the gigantic shift in the means of
> production has been sufficiently recognized by us. For we have to see our case
> in the light of technological history…The architect is in a very real danger
> of losing his grip in competition with the engineer, the scientist and the builder
> unless he adjusts his attitude and aims to meet the new situation.[33]

Rather than resist the overtures of industry, Gropius, Mies van der Rohe, and their contemporaries embraced the opportunities and technical support offered to them by the aluminum industry. The result was described repeatedly in various industry publications and press releases as the dawning of the Age of Aluminum.

## The Problems of Pervasiveness

Thus far, this essay has described the convergence of design theory, industrial expansion, and product development to account for aluminum's pervasive presence in the built environment. The account provided above could be expanded and refined to include the contributions of architects like Buckminster Fuller to the development of a design ideology based on the space-time-weight mantra. Fuller's Dymaxion House of 1927, for example, provides obvious parallels to the story of Kocher and Frey's experimental Aluminaire House. And although later European developments received little attention here, the career of the French architect Jean Prouvé could be cited as another important case study.[34] Or one could move forward in time and catalogue significant buildings erected in recent decades that continue the story of aluminum. But an account that focuses exclusively on aluminum's success in making the transition from precious to pervasive misses an important aspect of the story: resistance to the new architecture.

Despite the concerted efforts of modern architects like Albert Frey and Buckminster Fuller, American home buyers resisted the large-scale introduction of homes radically reconfigured as industrialized artifacts conducive to mass production.[35] And although the decades following World War II are often described as

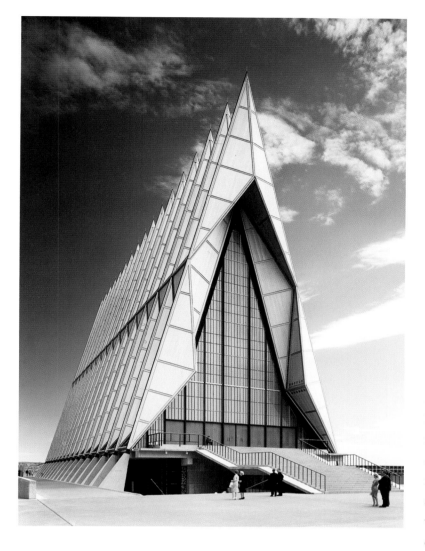

marking the triumphal ascendency of modernism in architecture, cultural resistance to bold new designs continued to surface from time to time in fields other than residential architecture. In July 1954, the architectural firm of Skidmore, Owings & Merrill received the commission for the design of the new United States Air Force Academy in Colorado Springs.[36] S.O.M. envisioned a campus of uncompromisingly modern buildings and employed aluminum, in the form of extruded sections, in the design of the curtain walls. In a dramatic departure from the rectilinear nature of most of the campus buildings, S.O.M.'s Walter Netsch designed a crystalline multispired chapel (fig. 115). Netsch created a space frame consisting of tetrahedrons clad with aluminum panels and narrow bands of stained glass. When, in the 1890s, Raffaele Ingami used aluminum for the roof of San Gioacchino in Rome, it failed to arouse any resistance. Again, in 1926, Henry Hornbostel's design for an aluminum spire for the Smithfield United Church in Pittsburgh failed to provoke strong opposition. In each case, aluminum served as a substitute for, rather than a challenge to, more traditional ways of building. Such was not the case for the Air Force Academy, however, and the design attracted numerous critics.[37] By the mid-1950s, aluminum was firmly identified in the popular consciousness as a quintessentially modern material, and in building types (like churches and homes) for which the public's commitment to modernism was still less than total, aluminum could be the target of criticism as well as praise.

Resistance to aluminum came not only from cultural conservatives; the success of aluminum increasingly made it the target of other industrial and trade groups with a stake in the building industry. The Allied Masonry Council, an organization representing the interests of brick, stone, and tile manufacturers and contractors, hired a Washington lobbyist to try to force changes in the metal and glass vocabulary of the Air Force Academy and, in congressional hearings related to appropriations for the Academy, the S.O.M. design was attacked by one congressman as "a monument to experimental materials."[38] Although the Air Force Academy ultimately was executed as designed by S.O.M., opposition by contractors, labor unions, and the suppliers of traditional materials did thwart some efforts to introduce newer materials and building technologies across the United States.[39]

In 1956, reflecting on his own enthusiasm early in the 20th century for industrialized prefabricated building systems, Walter Gropius remarked:

Figure 115.
**Skidmore, Owings & Merrill**
(American, 1936–present)
Lead architect: Walter Netsch
(American, b. 1920)
United States Air Force Academy
Chapel, Colorado Springs, 1956–62

At that time I still thought that, within a few years, everybody would accept this new industrial building method, but today I see how slowly such a process of change has gone because the inertia of the human heart is too great. Particularly in a period where everything has changed, not only the methods of production, but also our thinking, man tends to cling to the visible things he has inherited from his grandpa.[40]

Whether it was due to the "inertia of the human heart" on the part of home buyers or the attempts of participants in the construction industry to protect existing skills, markets, and trades, resistance to possible applications is an important part of the aluminum story. Examples of resistance to aluminum suggest that its history is far from simple. Explanations of aluminum's pervasiveness in the modern world that rely on deterministic frameworks—for instance, the inevitability of certain patterns and systems emerging once new technologies or materials have been invented—fail to convey the complexity of the aluminum story.

Aluminum presented architects with an unusual problem. "The danger with aluminum," Mies van der Rohe observed in 1956, "is that you can do with it what you like; that it has no real limitations."[41] When Henry Hornbostel designed the cast aluminum spire of the Smithfield United Church, he chose to emulate the traditional forms of filigree metalwork. His substitution of one lighter metal for another represents an early stage in the history of aluminum. Once architects accepted Otto Wagner's argument that "the sole departure point for our artistic work can only be modern life," the foundations for aluminum's emergence as a major architectural metal were laid. But, as this essay has tried to demonstrate, the story of aluminum's success as an architectural metal cannot be told exclusively from the architects' point of view. The development of aluminum applications in architecture shaped and was shaped by the aspirations of various parties. Some historians of technology describe a historical process in which new materials and systems are introduced, adjust to conditions, exploit opportunities, and gather momentum until success *seems* inevitable.[42] Aluminum provides an excellent case study of the phenomenon of technological momentum. The momentum that propelled aluminum to its prominent place among 20th-century building materials was the result of the synergy generated in the interplay of three different historical "actors." Aluminum producers in search of markets in a competitive environment constitute one historical agent in this story. Architects in search of industrialized materials and systems adaptable both to building needs and to a vision of modernity represent a second. And the third comprised inventors like Francis Plym, who developed new products and building systems that capitalized on the advantages offered by aluminum. If the 20th century can be characterized fairly as the Age of Aluminum, then it is worthwhile to reflect on the fact that the Age of Aluminum didn't just happen, it emerged when and as it did because of the combined efforts of daring visionaries, hardheaded industrialists, and enterprising inventors.

## Notes

1. J. Eveleth Griffith, ed., *History of the Washington Monument from Its Inception to Its Completion and Dedication* (Holyoke, Mass.: J. Eveleth Griffith, Printer, 1885), 11.

2. Carroll L. V. Meeks, *Italian Architecture, 1750–1914* (New Haven: Yale University Press, 1966), 386–94.

3. See Margot Gayle, David W. Look, and John G. Waite, *Metals in America's Historic Buildings*, 2d ed. (Washington, D.C.: U.S. Department of the Interior, National Park Service, 1992), 84.

4. "An Architect's Opinion of Aluminum," *The Aluminum World* 5 (October 1898): 8.

5. Otto Wagner, *Modern Architecture*, trans. Harry Francis Mallgrave (Santa Monica, Calif.: Getty Center for the History of Art and the Humanities, 1988), 60.

6. Ibid., 78.

7. Ludwig Mies van der Rohe, "Industrialized Building," *G* (June 10, 1924); translated and republished in Ulrich Conrads, ed., *Programs and Manifestoes on 20th-Century Architecture* (Cambridge, Mass.: MIT Press, 1970), 81–82.

8. Frank Lloyd Wright, "In the Cause of Architecture. VIII: Sheet Metal and the Modern Instance," *Architectural Record* 64, no. 4 (October 1928): 334.

9. Ibid., 335.

10. An account of Plym's invention along with a history of the Kawneer Company appears in Thomas Stritch, *The Kawneer Story* (Niles, Mich.: Kawneer Company, 1956), esp. 7–21.

11. *Kawneer Solid Steel Windows, Catalogue F* (Niles, Mich.: Kawneer Mfg. Co., 1915), 3.

12. See, for example, Charles W. Killam, "Modern Design as Influenced by Modern Materials," *Architectural Forum* 53, no. 1 (July 1930): 39–42; Douglas B. Hobbs, "Aluminum in Architecture," *Architectural Forum* 53, no. 2 (August 1930): 255–59; David Emerson, "Metals and Alloys," *Pencil Points* 12, no. 3 (March 1931): 239–40; John Cushman Fistere, "The Use of White Metals," *Architectural Forum* 55, no. 2 (August 1931): 233–40; K. Lönberg-Holm, "Aluminum," *Architectural Record* 71, no. 1 (January 1932): 64–65. For a review of the history of other modern metals, see Thomas C. Jester, ed., *Twentieth-Century Building Materials: History and Conservation* (New York: McGraw-Hill, 1995).

13. Harold W. Vader, "Aluminum in Architecture," *Architectural Record* (December 1931): 461–62.

14. Neil Jackson, "Aluminaire House, USA (Kocher and Frey)," in *Modern Movement Heritage*, ed. Allen Cunningham (London: E & FN Spon, 1998), 136–44.

15. Joseph Rosa, *Albert Frey, Architect* (New York: Rizzoli, 1990). Rosa's monograph includes an extensive bibliography for both Frey and the Aluminaire House.

16. A. Lawrence Kocher and Albert Frey, "Windows," *Architectural Record* (February 1931): 127–37.

17. Jackson, "Aluminaire House, USA," 137.

18. Le Corbusier's "Five Points of the New Architecture" were (1) pilotis (slender columns), (2) a roof terrace, (3) a free plan, (4) horizontal windows, and (5) a free facade. For a discussion of Le Corbusier's Five Points, see William J. R. Curtis, *Modern Architecture since 1900*, 3d ed. (Upper Saddle River, N.J.: Prentice Hall, 1996), 176–77.

19. For a discussion of experimental homes in the period, see Brian Horrigan, "The Home of Tomorrow, 1927–1945," in *Imagining Tomorrow: History, Technology, and the American Future*, ed. Joseph J. Corn (Cambridge, Mass.: MIT Press, 1986), 137–63.

20. For a complete description of this and other contemporary designs by Frey, see Rosa, *Albert Frey*, chapter 3, "America, the West Coast, 1939–1955."

21. Sidney K. Robinson, "The Postwar Modern House in Chicago," in *Chicago Architecture and Design, 1923–1993*, ed. John Zukowsky (Chicago: Art Institute, 1993), 201.

22. Dennis Doordan, "Promoting Aluminum: Designers and the American Aluminum Industry," *Design Issues* 9, no. 2 (Spring 1993): 44–50.

23. For more information on the Empire State Building, see Carol Willis, ed., *Building the Empire State* (New York: W.W. Norton, 1998), and John Tauranac, *The Empire State Building: The Making of a Landmark* (New York: Scribner, 1995).

24. R. H. Shreve, "The Empire State Building II: The Window-Spandrel-Wall Detail and Its Relation to Building Progress," *Architectural Forum* 53, no. 1 (July 1930): 101.

25. *Aluminum on the Skyline* (Pittsburgh: Alcoa, 1954), 3.

26. John Peter and Paul Weidlinger, *Aluminum in Modern Architecture*, 2 vols. (Louisville, Ky.: Reynolds Metals Company, 1956).

27. "An International Award," Reynolds Metals Company press release, November 12, 1956.

28. Gaylord Harnwell, "Contemporary Building," Reynolds Metals Company press release, November 12, 1956.

29. In May 1943, for example, *Architectural Forum* published a special issue with designs for twenty-three different building types conceived as part of the plan for a hypothetical town of 70,000; the theoretical exercise optimistically looked forward to the resumption of large-scale building activity after the war. See "New Buildings for 194X," *Architectural Forum* 78, no. 5 (May 1943).

30. "Store Fronts of Tomorrow. A New Pencil Points–Kawneer Competition," *New Pencil Points* 23, no. 10 (October 1942): 34.

31. "Store Fronts of Tomorrow," *New Pencil Points* 24, no. 2 (February 1943): 30–41.

32. *Store Fronts of Tomorrow by Kawneer* (Niles, Mich.: Kawneer Company Brochure, 1943), n.p.

33. Walter Gropius, *Scope of Total Architecture* (New York: Harper, 1955), 73. Gropius was hardly alone in his call for the architectural profession to reconfigure its relationship with American industry. Already in the early 1930s, American architects were convinced that traditional conceptions of professional practice were increasingly outmoded; see, for example, E. D. Pierre, "We Must Become Part of the Building Industry," *American Architect* 139 (May 1931): 64.

34. For an account of Prouvé's experiments with prefabricated building systems, see Benedikt Huber and Jean-Claude Steinegger, eds., *Jean Prouvé. Prefabrication: Structures and Elements* (New York: Praeger, 1971).

35. Trailer homes and recreational vehicles, which fall outside the scope of this essay, constitute possible exceptions to this generalization.

36. For documentation of the Air Force Academy story, see Robert Bruegmann, ed., *Modernism at Mid-Century: The Architecture of the United States Air Force Academy* (Chicago: University of Chicago Press, 1994).

37. For the controversy surrounding the Air Force Chapel, see Sheri Olson, "Lauded and Maligned: The Cadet Chapel," in Bruegmann, ed., *Modernism at Mid-Century*, 156–67.

38. Opposition to the Air Force Academy design, including remarks by Representative John Fogarty (D–R.I.), is described in detail in Bruegmann, ed., *Modernism at Mid-Century*, 43–44.

39. See, for example, Robinson's discussion of labor-union resistance to new service systems in domestic construction in the Midwest; Robinson, "The Postwar Modern House in Chicago," 202–03.

40. Walter Gropius, "The Future of Aluminum in Modern Architecture," in Peter and Weidlinger, eds., *Aluminum in Modern Architecture*, vol. 1: 228.

41. Ludwig Mies van der Rohe, "The Future of Aluminum in Modern Architecture," in Peter and Weidlinger, eds., *Aluminum in Modern Architecture*, vol. 1: 248.

42. See examples in Thomas Hughes, "Technological Momentum," and Thomas Misa, "Retrieving Sociotechnical Change from Technological Determinism," in *Does Technology Drive History? The Dilemma of Technological Determinism*, ed. Merritt Roe Smith and Leo Marx (Cambridge, Mass.: MIT Press, 1994).

SILVER & CO.

Sole Manufacturers of

"BROOKLYN" PURE ALUMINUM GOODS.

304-314 HEWES STREET,

BROOKLYN, N. Y.

SILVER & CO'S.

"ALUMINUM CUP"
JUICE EXTRACTOR.

SILVER & CO'S
PURE
ALUMINUM
COMBINATION
LEMON JUICE
EXTRACTOR
SHAKER
AND
MIXER.

Cookware to Cocktail Shakers:

The Domestication of Aluminum in the United States, 1900–1939

Penny Sparke

Much has been written about the dramatic changes in women's lives that have taken place in the 20th century. Those accounts that focus on the domestic context inevitably highlight the role of technology. Ruth Schwartz Cowan and Judy Wajcman, for example, have both written about the domestic industrial revolution and its dramatic effects on women.[1] They have described the impact of new power sources— gas and electricity—and the emergence of modern domestic appliances, stressing the new behaviors that grew from these developments and the ultimate futility of attempts to transfer industrial rationalism into the home. What they and others have overlooked, however, is the material face of these transformations, that is, the role of new industrial materials as both reflections and determinants of social and cultural change.

Here, I will trace the impact of aluminum on women's lives during the first four decades of the 20th century, decoding the shifting meanings of the domestic aluminum object in the context of consumption and use. Central questions are: how the aluminum object assumed the cultural values that were constructed for it through advertising and sales strategies; and how those values, and the objects expressing them, were negotiated as they entered the domestic sphere. The aluminum industry has operated on a mass-production basis since the early years of the 20th century. Consequently, there has been a long-standing need for an equally large-scale consumption of aluminum objects. In the following discussion, I will attempt to ascertain the extent to which aluminum objects were either welcomed or resisted in the home. Were they perceived as transmitters of values toward which the early 20th-century housewife aspired, or did she have to be won over to the virtues of aluminum to complete the production/consumption cycle? Did aluminum objects, in their modern industrial guise, play a key role in constructing an image of modernity for women in the early 20th century?

The cultural context of women's lives that frames this inquiry is the scientific model of the housewife, in which a "rational" preoccupation with issues of efficiency, health, and safety was uppermost. This model was gradually joined—and subsequently displaced—by "Mrs. Consumer," who put more value on the "irrational," emotional aspects of domesticity: consumer desire, leisure, nurturing, and familial happiness.[2] In search of a mass market for its goods in the first decades of the

Figure 116.
Silver & Co. advertisement
*The Aluminum World* 1
(July 1895): 200

20th century, the aluminum industry had to contend with these shifting cultural sands. Arguably, in doing so, it became an agent of cultural change rather than merely a passive reflection of it.

## From Invention to Application: The Birth of the Aluminum Cookware Industry

Aluminum was discovered in Europe in the 1840s. It remained an elite novelty material, however, until such time as it could be produced at a much lower cost. This was achieved in 1886, on both sides of the Atlantic, by the Frenchman Paul Héroult and the Ohio-born Charles Martin Hall, who transformed the new metal into a more widely accessible "solution in search of a problem." In the following decade, the need for an application in the American context led to countless thwarted projects, among the least successful a hairfanning machine, a harp, and a ventilating hatband.[3]

The early years of the aluminum object aimed at the domestic market were dominated by the emergence of aluminum kitchenware. A range of cooking utensils, among them kettles, pots and pans, tea- and coffeepots, waffle irons, griddles, and egg boilers, utilized significant amounts of the available metal and offered the housewife a viable substitute for her existing kitchen utensils of cast iron, tin, or porcelain enamel (fig. 116).

The 1890s were transitional years in which cast and stamped aluminum cookware began to be manufactured in an atmosphere of both consumer skepticism and intense industry competition. The decade witnessed a scramble among a few firms to capture the still-modest market for these goods. In 1891–92, the powerful Pittsburgh Reduction Company (formed in 1888 and eventually renamed Alcoa) entered into a wrangle with the Griswold Manufacturing Company over who would undertake the manufacture of aluminum products. Griswold commissioned the fabrication of cast aluminum teakettles from the Pittsburgh Reduction Company (lending the latter a molder to perform the job), but finally moved into manufacture itself in 1893 (fig. 119).[4] The Illinois Pure Aluminum Company unveiled its stamped aluminum cookware in 1892, alongside hunting boats, while the Wagner Ware Manufacturing Company exhibited its cast aluminum teakettles in the first year of their production at the 1893 World's Columbian Exposition in Chicago.[5] A certain Sara T. Rorer, who was in charge of the Model Kitchen at that event, was enthusiastic about the new cookware, claiming that it did not scorch food.[6] A tremendous sense of optimism surrounded these achievements, to which an 1897 letter from a member of Griswold to the Pittsburgh Reduction Company bears witness: "Judging from the results of our work we have no question but what these goods are going to increase in popularity, and become in great demand as time goes along until no first class house-keeper will be satisfied without her full set of aluminum kitchenware."[7]

The writer's enthusiasm proved to be premature, however, as it was soon apparent that more work would have to be done to capture the large domestic market

(opposite)
Figure 117.
(left) **John Gordon Rideout,**
designer (American, d. 1951)
**Harold L. van Doren,** designer
(American, 1895–1957)
Wagner Ware Manufacturing
Company, now owned by
General Housewares Corporation,
manufacturer (American,
1891–present)
Magnalite teakettle, ca. 1940
aluminum and nickel alloy
and lacquered wood
cat. 3.63

(center) Sears, Roebuck & Co.
(American, 1893–present)
Maid of Honor teakettle, ca. 1930
aluminum and plastic
cat. 3.64

(right) Wagner Ware Manufacturing
Company, now owned by
General Housewares Corporation
(American, 1891–present)
Grand Prize teakettle, 1916
aluminum and wood
cat. 3.65

needed to make the new aluminum cookware industry a profitable one. Aluminum goods manufactured in the second half of the 1890s were of poor quality, and housewives were not persuaded by claims of their lightness and brightness, absence of seams, speed in conducting heat, and ability to retain heat. In fact, consumers distrusted the metal's light weight, and were concerned about its tendency to discolor. The expansion of the home preservation of fruits and vegetables at this time—a mark of the efficient homemaker, who increasingly had to prove her worth—meant a large market for preserving pans for which lightness was appreciated but discoloration was not. Moreover, the cost of aluminum ware was still relatively high. In the years immediately after 1900, therefore, the need to find subtler ways of promoting aluminum goods was paramount, and many minds were dedicated to the project.

### The Power of Persuasion: Promoting and Selling Aluminum Cookware

Between 1900 and 1918, the market for aluminum cookware experienced enormous growth, such that by the end of the period it had found a place in the material culture of everyday American domestic life (fig. 117). This was primarily the result

Figure 118.
Griswold Manufacturing Company
display, Paris Exposition, 1900
*The Aluminum World* 6
(March 1900): 116

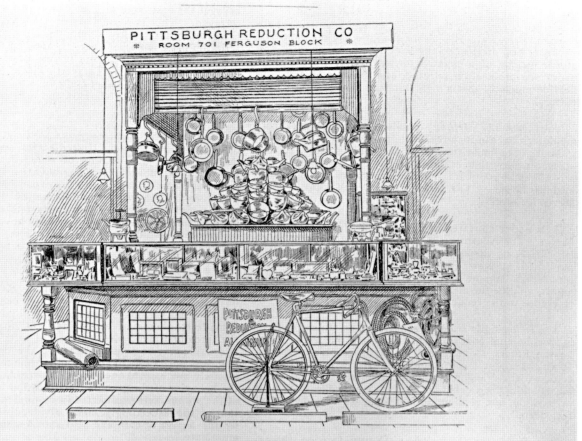

**ALUMINUM EXHIBIT AT THE PITTSBURG EXPOSITION**

THIS ENGRAVING IS AN EXACT COPY OF A PHOTOGRAPH TAKEN FOR THE ALUMINUM WORLD, AND SHOWS THE MANY USES OF ALUMINUM IN THE ARTS.

Figure 119.
Pittsburgh Reduction Company aluminum exhibit, Pittsburgh Exposition, 1894
*The Aluminum World* 1
(December 1894): 47

of two forces: first, the determination of the industry to find new ways to promote and sell its goods and, second, the ideological convergence of the rhetoric surrounding the promotion of aluminum kitchenware, which stressed its labor-saving properties, with the value system of scientific management in the home (fig. 118).

A high level of ingenuity was required of advertisers, marketing men, and salesmen to get a critical mass of aluminum goods into the kitchen. The restaurant and hotel trades were expanding in the early years of the new century, but in order to justify the high investment costs involved in making aluminum cookware, a large domestic market was needed as well. In 1900, less than 5 percent of all aluminum produced was used for cooking utensils. On the other hand, the raw material was becoming more accessible, and the first few years after 1900 saw the increasing power of the Pittsburgh Reduction Company, which controlled the price of raw aluminum. The company formed its own selling operation—the Aluminum Cooking Utensil Company—in 1901, and in 1903 founded the Wear-Ever brand of aluminum utensils, which would become a household name within little more than a decade.

A key factor shaping the strategies adopted at this time to promote aluminum cookware was the transformation of women's domestic lives. Back in 1895, a writer in *The Aluminum World* had noted the importance of the new goods in "these days of light housekeeping and amateur cookery,"[8] indicating that the role of the housewife was changing. The principal reason for this change was the so-called servant problem—that is, the diminishing availability of live-in domestic help—which resulted in the middle-class housewife assuming more responsibility for domestic work. Added to this was the growing impact of "germ theory," which claimed that the spread of disease was directly linked to household dust and dirt. An obsession with "health"—the family's and, by implication, the nation's—was pervasive in household manuals of the time, which exhorted the housewife, or her newfound "professional" equivalent, the "household manager," to prioritize cleanliness and to emulate the scientist in the rational management of the home. This alignment with science reached a peak with the publication, in the *Ladies' Home Journal* in 1913, of Christine Frederick's series of essays collectively entitled *The New Housekeeping.*[9] By that time, aluminum cookware had found a place in the modern American kitchen as a weapon in the household manager's arsenal.

This happy coincidence was not achieved, however, without enormous effort on the part of manufacturers, nor without the deliberate construction of a set of meanings for aluminum cookware, mediated through advertising, marketing, and sales strategies devised in the period leading up to World War I. An important and recurring selling technique that allied housework with the activity of the scientist was the use of laboratory-like "experiments" or "tests" set up to show potential consumers the quantifiable benefits of aluminum cookware. In one such experiment, undertaken in 1903 in Wanamaker's department store in Philadelphia, apples were cooked in a Wear-Ever aluminum pan without being stirred and, as demonstrated to incredulous customers, without sticking to the bottom.[10] A decade later, the Aluminum Cooking Utensil Company published a pamphlet entitled *The Wear-Ever Kitchen,* which addressed "progressive housewives [who] are now studying as never before household economy" and presented them with a series of tests showing how

> …food does not scorch as readily in "Wear-Ever" utensils as in other utensils, how fuel may be saved by using them, why the utensils are not injured if food is accidentally burned in them, how they may be cleaned, how they may be cared for so as to give you life-long service.[11]

Empirical, and therefore "irrefutable," evidence of aluminum's superiority was provided by such tests, which included holding an aluminum pan and an enamel pan over a flame to see which one the demonstrator dropped first; cooking a pint of tomatoes to burning point in an aluminum pan and showing that the pan was not harmed; and throwing a cup of ice water into a hot aluminum pan

Figure 120.
"Test Number One," from *The Wear-Ever Kitchen* (New Kensington, Pa.: Aluminum Cooking Utensil Company, ca. 1912), 5

## Test Number One

Hold by the edges an aluminum pan and an enamel pan of the same size so that the sides of the pan opposite the hands come in contact with a small flame of fire. You will drop the aluminum pan first.

HEAT passes through aluminum twice as fast as through tin and three times as fast as through iron or steel. Heat does not collect in the bottom of the utensil, but spreads rapidly and evenly throughout all parts of it—thus preventing local overheating, which is the cause of burning. This explains why bubbles may rise more quickly from the bottom of an enamel utensil than from an aluminum utensil. Quiet, even heat, not violent motion, cooks food.

Aluminum will store up more heat and retain it longer than any other metal.

Less heat, therefore, is needed to cook food in "**Wear-Ever**" utensils than ordinarily is used.

Figure 121.
The Modern Kitchen display window
*House Furnishing Review* 29, no. 4
(October 1908): 31

to show that it would not crack (fig. 120). The strongly "academic" tone of these operations was reinforced by testimony provided for Wear-Ever by "Jos. W. Richards, Ph.D., Professor of Metallurgy, Lehigh University, Bethlehem, Pa." at the back of the pamphlet, in which the professor claimed he had used his aluminum utensils daily for eighteen years.[12] "Scientific" tests were performed widely through the first decade of the century in stores and homes, and at trade fairs and exhibitions, to persuade the housewife that aluminum was *the* efficient, labor-saving material for modern cookware. Its alliance with science suggested that it was a material of the "modern age," a product of experimentation and of the technological advances associated with the process of industrialization. As such, it could "modernize" the home, hitherto a bastion of traditional values but fast becoming redefined in the new century. This emphasis on the "modern" meaning of aluminum cookware was not yet apparent in the visual styles used for advertisements, which were either historicist or utilitarian in flavor, but it was present, nonetheless, in rhetoric about the home invoked in popular journals of the day. *House Furnishing Review*, for example, carried an article in 1908 entitled "The Modern Kitchen," which described a new, very modern-looking retail outlet of that name in New York, run by Miss Helen M. Logan, whose aim was to provide "Mrs. Newlywed" with all her household utensils (fig. 121).[13] By implication, the concept of "modernity" was

thought to have a strong appeal for America's young female consumers in the process of setting up house.

In the United States at this time, to be "modern" was not only linked with individual aspiration and social status. It was also synonymous with being "American" and, as a result, aluminum could be seen as a potential messenger not only of the "rational household" but also of nationhood. This was confirmed by a 1911 advertisement for the National Aluminum Works boasting a large American eagle surrounded by aluminum pots and pans (fig. 122).

Promotional material for aluminum cookware flattered potential customers by addressing them as intelligent and capable of understanding detailed technical data with a scientific orientation. *Aluminum Facts*, published by the Aluminum Cooking Utensil Company in 1906, presented customers with tests to perform at home and information about the metal's durability, safety (fear of being poisoned by aluminum had been expressed often by housewives), health-promoting properties (grease was not needed when baking cakes in aluminum), heat-retaining capacity, and close molecular structure (in other words, it was nonporous) (fig. 123).[14]

From early in the century, aluminum goods were sold through retail outlets, including department stores, and door-to-door by salesmen. A strong emphasis was placed on the training of traveling salesmen. In 1907, the Aluminum Cooking Utensil Company issued its *Instructions to Salesmen*, which revealed the practical and psychological training these men were given. Canvassers were expected to carry

Figure 122.
The National Aluminum Works advertisement
*House Furnishing Review* 34, no. 1
(January 1911): 22–23

## LIGHTNESS

**Y**OU have decided before reading this far that you *must* have aluminum utensils. But read this anyway and you will become an enthusiast over aluminum.

There are many women who may not consider the wonderful lightness of aluminum as a cardinal virtue. They have gone on using back-breaking iron pots and skillets until they like to consider themselves martyrs. But others there are who consider health and comfort and will at once adopt any means to insure receiving them.

Do you know that aluminum is but one-fourth the weight of silver? That a block of aluminum, as tough and almost as strong as steel, will weigh little more than a block of wood of the same size?

As for comparing the weight of an aluminum kettle with the weight of an iron kettle of equal capacity, you wouldn't believe it if we told you the difference in weight.

A gallon of water is heavy in itself; add half as much again to that weight by the iron pot in which you lift it, and the weight becomes dangerous for the average woman to have to handle.

A gallon capacity aluminum kettle when empty is so light that you scarcely know you have hold of it.

You husbands with tired wives— isn't aluminum worth trying?

9

Figure 123.
"Lightness," from *Aluminum Facts* (New Kensington, Pa.: Aluminum Cooking Utensil Company, 1906), 9

samples and test equipment with them (the tomato test was especially popular), and to plan their sales campaign as if it were a military operation. The aim was to penetrate the homes of the influential women in a given town—details about whom could be gathered from the town hall and the postmaster, among other sources—and to persuade them of the benefits of aluminum cookware through practical demonstrations. As the booklet explained: "Do not show your samples at the door. Get inside. Do this by interesting your prospective customer and thereby receiving an invitation into the house, rather than by insisting upon admittance without first exciting interest."[15] This was a "hard" selling technique that clearly reaped rewards, as the level of aluminum consumption rose considerably during the first decade of the century.

Door-to-door sales was dominated by men, largely because of the lifestyle that came with the occupation. But selling aluminum cookware was also being done increasingly by women to women. Nowhere was this more evident than in the store demonstrations that breathed life into inanimate goods and, as a consequence, brought them one step nearer to the home. January 1908, for example, was dedicated to "daily lectures and demonstrations on the art of preparing, cooking and serving dainty foods" by Mrs. F. Violet Black in the household and kitchen department of the William Barr Dry Goods Company store in St. Louis. The intention was "to instruct and educate the housewife in the better performance of her duties and bring to her attention many labor-saving devices needed in the modern kitchen."[16] It was a clear message, one which bridged the gap between educating and selling, and served to ally the two in the minds of the new "household managers" such that consumption was perceived as part of the "professional" activity of being a housewife.

Advertising was equally hard-hitting and, once again, stressed the quantifiable, scientific "truths" about this new material believed to appeal to the "modern woman." In the years leading up to World War I, the "woman-to-woman" approach found favor in this arena as well. Men ceased to be the sole authorities whose testimony legitimized the purchase of aluminum cookware items. In their place, advertising agencies created the mythical "new women" who, like the home economists and female demonstrators, were on the consumer's side and could be trusted. The first move in this direction was the inclusion of female mannequins dressed as housewives in aprons, ensconced in shop window displays, providing "someone" with whom the consumer could identify. Prior to this, window displays had consisted of enormous piles of utensils with no reference to use or to the gender of the user. From about 1911, the user was depicted as female and displays looked less like factory storerooms and more like lived-in domestic environments. In 1912, for example, a store owned by William H. Frear & Co. won a prize for its display of Wear-Ever

products presented in a kitchen setting complete with furniture and striped window curtains (fig. 124).

By 1911, Wear-Ever was being advertised in a vast range of women's magazines, from *Women's Home Companion* to the *Ladies' Home Journal* to *Good Housekeeping* to *The Delineator*. At about the same time, illustrations of housewives appeared in magazine advertisements for aluminum cookware, shown using, cleaning, or testing their purchases, often with a soft smile of satisfaction. Children appeared as well, reinforcing the idea that it was the housewife herself rather than the servant who was being depicted. In an advertisement in a 1914 *House Furnishing Review*, a bride, dressed in her wedding gown, could be seen gazing lovingly at a set of aluminum utensils (fig. 125). A little later, the female mannequin and abstract illustrated figure were incorporated by the Massillon Aluminum Company of Ohio in the figure of "Betty Bright," who was shown in advertisements with her head emerging from an aluminum saucepan (fig. 126). Betty was *the* modern housewife, rather than simply *a* modern housewife, a specific individual to whom all other aspiring modern housewives could relate directly.

While displays of aluminum goods in stores, at trade fairs and national exhibitions, and in magazine advertisements became more sophisticated and consumer-oriented in the first decade and a half of the century, the appearance of the goods themselves did not mirror this sophistication. Shapes remained simple, utilitarian, and technologically determined, and the only modifications related to function. More and more multipurpose utensils were created to increase the speed and efficiency of food preparation. The visual appearance of the goods themselves was rarely mentioned, although the bright, silvery surface was frequently described

Figure 124.
William H. Frear & Co.'s prize-winning display of Wear-Ever aluminum
*House Furnishing Review* 37, no. 4 (October 1912): 6

as "beautiful" and as an asset to any display. Forms changed only to accommodate increased efficiency and, when used, the word "design" referred to the performance of the products rather than their appearance. In an advertisement from 1911, for example, promoting the "best designs," the goods illustrated exhibit the familiar simple forms, although several—a combination egg poacher and a four-element "heat container" among them—had multiple components and were clearly ingeniously conceived. Practicality, utility, and efficiency were the order of the day. In themselves, these qualities suggested the "modernity" that was associated with aluminum goods and that appealed so strongly to their consumers.

The key meanings of the modern aluminum object were firmly established, therefore, in the years leading up to World War I. These meanings were visible in the growing feminization of the promotional material; in the scientific emphasis of promotional and sales strategies; and in the close alliance of the products with the ideological base-lines of feminine modernity as it emerged in these years. As well as reflecting these cultural trends, the promotional material was itself helping to define women as consumers of goods for the domestic sphere, and addressing them more and more directly. The manufacturing companies, in conjunction with advertising agencies, the retail sector, and exhibition culture, played a very active role in this process. At this time, however, the goods themselves were innocent carriers of these messages rather than consciously contrived visual symbols of them. This would change during the postwar years.

### The Housewife's Dream: From Needs to Wants

Figure 125.
Wear-Ever advertisement
*House Furnishing Review* 40, no. 5
(May 1914): 2

Figure 126.
Massillon Aluminum Company
advertisement
*House Furnishing Review* 43, no. 1
(July 1915): 11

The years between the end of the war and the late 1920s witnessed a significant shift in the meaning of the aluminum object in the context of women's domestic lives. In essence, it moved from being a new item in the kitchen, accepted on the grounds of its proven benefits in terms of efficiency, to become a highly popular "object of desire," one which was still believed to improve the effective running of the modern home but which was also an important material symbol of that progressive ideal. As a result, the aluminum object became increasingly integrated into the ideology of domesticity that dominated American feminine culture during the 1920s.

Moreover, the housewife's responsibility expanded from running the home like an efficient machine—as time was (supposedly) freed by the advent of labor-saving goods and appliances—to considering the emotional and cultural requirements of modern family life. In the 1920s, housewives found themselves having to address notions such as comfort, beauty, and happiness in the domestic sphere as well as its efficient functioning. Whether this was the result of women themselves redefining their role, or of manufacturers and retailers finding new ways of appealing to them, is a subject for conjecture.

This change in domestic ideology occurred in a context of expanded manufacturing, and the production of aluminum cookware was no exception. The number of companies making and selling aluminum kitchenware increased threefold between 1914 and 1920. About forty such companies were operating in 1924. New names included "Mirro" (Aluminum Goods Manufacturing Company), "American Maid" (Popular Aluminum Cooking Utensils), the Club Aluminum Utensil Company, MonarCast Aluminum Cooking Utensils, and Aluminum Products Company, while the established names of Wear-Ever, Wagner, and Griswold continued to flourish. Inevitably, some fell by the wayside as the economic climate became tougher toward the end of the 1920s but, in general, the industry went from strength to strength. The Aluminum Wares Association launched a national campaign to increase sales by $50 million in the latter half of the decade.

Industrial expansion was matched by widespread acceptance of aluminum kitchenware, which had lost its "shock of the new" quality and had largely overcome anxieties about its safety. Promoters were still careful to make clear that scientific tests had proven that aluminum would not contaminate food, indicating that such reassurances were still necessary. Overall, however, aluminum had become the accepted material for many of the kitchen utensils required by the modern housewife. Once again, the role of promotion in creating this situation should not be underestimated.

The strategies used in promotional material from this period show clearly how the popularity of aluminum cookware was achieved and sustained. Key to these was a shift in the way such items were represented, in terms of the shifting definition of feminine domesticity. While scientism was still invoked in the selling of aluminum cookware, there was a growing suggestion as well that these goods could play a more subjective role in the life of the modern housewife. This was achieved by a deepening alignment of aluminum utensils with women and their new lifestyle; by the emergence of a class of utensils that could move out of the rational environment of the kitchen and into the dining room; and by a growing emphasis on utility *and* beauty as the defining characteristics of the aluminum object.

Wartime exigencies had demanded thrift and efficiency, and in the early postwar years these values were deployed in American advertising rhetoric. Aluminum fit easily into this rhetoric and all the qualities of the material established in the prewar years—labor-saving, heat-conducting, heat-retaining, noncorrosive, seamless, easily cleaned, light—were reconfirmed as reasons why housewives should welcome it into their kitchens. The concept of the multipurpose aluminum cooking object

**An Aluminum Miracle— 18 Utensils In One**

Patent No. 53879.
October 7, 1919.

*"It certainly makes cooking easy!"*

also emerged at this time, justified on grounds of thrift and efficiency. An eighteen-in-one set was offered in 1920, for instance, which could "boil, bake, stew, steam, poach, and strain" and which saved stove space, time, and fuel (fig. 127).

Advertisers went further down this rational road in support of the "professional" housewife whose role extended far beyond mere efficiency. An article in *House & Garden* in 1920 called this new feminine professional "Mrs. Man-of-the-House" and suggested that she could define herself as an equal to her husband through the manner in which she purchased her kitchen utensils:

> But what about Mrs. Man-of-the-House? How about her equipment? How about making her business pay? Foolish, you say. Not a bit of it! Her business is making an attractive home. A home to which Father returns in the evening, tired after the cares of a business day. It behooves her then to put this business of hers on a paying basis, so that Father will have no excuse for feeling a superior creature.[17]

This Victorian "separation of the spheres" approach remained fundamental through the 1920s, but the duties of the housewife became more complex and, with the loss of servants, included both practical and emotional obligations. Aluminum cookware was frequently depicted in ads featuring women and children working together happily in the kitchen. Wear-Ever adopted this approach and from 1920 onward stressed the importance of the "well-appointed home" in which the kitchen was seen as "so important because of its influence on family health and happiness."[18]

The penetration of aluminum cookware advertising into modern feminine culture in the 1920s was accomplished through several strategies. Betty Bright made more frequent appearances in her role as a "mythical modern housewife" selling wares made by the Massillon Aluminum Company of Ohio (fig. 128). As she herself explained in first-person addresses to the consumer, "If I should see in a special sale

Figure 127.
Berkeley Sales Company
advertisement
*House & Garden* 37, no. 2
(February 1920): 112

Home-made
Apple Pie!

PIES baked in "Wear-Ever" aluminum pie pans taste as good as they look. Thick, delicious fillings of apples, peaches, pineapple, cherries or other fruits or berries in season! Crisp, tender, flaky, golden-brown crusts!

"Wear-Ever"
Aluminum Cooking Utensils

are preferred by women who are as particular about the utensils in which they cook food as they are about the dishes from which they serve it. "Wear-Ever" utensils are clean, bright and silver-like in their shining beauty.

"Wear-Ever" utensils are made in one piece from thick, hard sheet aluminum. No joints in which food can lodge. Cannot chip, rust or scale.

Replace utensils that wear out with utensils that "Wear-Ever"
Look for the "Wear-Ever" trade mark on the bottom of each utensil

The Aluminum Cooking Utensil Company, Dept. 36, New Kensington, Pa.
In Canada "Wear-Ever" utensils are made by Northern Aluminum Company, Limited, Toronto, Ont.

My Kitchen
"I love its 'homey,' orderly air and its immaculate cleanliness. Walls and ceiling are bright and spotless. My little stove beams like a pleased ebony god. My chairs and well-scoured table sit primly in their places like new scholars afraid to move. Sunshine falls in cheery squares through the crystal panes. My silverlike pots and pans wink back the dancing high-lights. They seem to say: 'Let's make play of work'!"

Edith Edwards

"The
Wear-Ever"
Specialty
Equipment

"Wear-Ever" gives to the Kitchen an atmosphere in keeping with the beautiful furnishings of the other rooms of the home.

such things as Dish Pans, Preserving Kettles, Roasters, Tea Kettles and Percolators, I'm afraid I simply couldn't resist the impulse to buy." Another strategy was to move the emphasis from the utilitarian cookware to the more sensorial appeal of the food cooked in it. Some advertisers simply added images of food to their press ads, which before had shown the somewhat stark utensils in isolation. As early as 1919, Wear-Ever had added large, mouthwatering images of ham and eggs, homemade apple pie, and apple pudding, accompanied by copy claiming that "Pies baked in 'Wear-Ever' aluminum pie pans taste as good as they look" (fig. 129).[19] This overt appeal to the senses opened the way for a new emotional relationship between the housewife and her aluminum cookware, and for a shift in the identity of the modern woman from that of scientist to consumer. This shift was apparent on a number of levels. In 1920, for example, *House Furnishing Review*, ever conscious of the need for new selling strategies to expand consumption levels, appointed Mrs. Edith H. Oliver as an editor to deal with the housewares section of the periodical so that a woman could talk directly to women in "woman's language."[20] In the same edition, a text in the magazine explained that "It is not only what she [the housewife] needs, that you must tempt her with, you must appeal to her desire for better things. She can get along with the old things and she will, unless you show her the better in something new."[21]

The emphasis on healthy and tasty food resulting from the choice of aluminum utensils was intensified by the publication of recipe books by aluminum cookware manufacturers, taking promotion one step further into the kitchen. A little brochure published by the Aluminum Cooking Utensil Company, which stressed the way in

Figure 128.
Massillon Aluminum Company advertisement
*House Furnishing Review* 59, no. 1 (July 1923): 58

Figure 129.
Wear-Ever advertisement
*House & Garden* 36, no. 4 (October 1919): 67

Figure 130.
Cover of *The "Wear-Ever" Specialty Equipment* (New Kensington, Pa.: Aluminum Cooking Utensil Company, n.d.)

which its "silverlike" kitchen utensils were the equivalent of the beautiful silver items "designed by artists," as well as the way in which work in the home was increasingly turning into a leisure activity, included recipes for, among other things, pot roast, escalloped apples, and poached or steamed eggs (fig. 130).[22] *Food Surprises from the Mirro Test Kitchen* showed how aluminum utensils could be used to cook roast beef and browned potatoes, a quick meat roll, or a ham shoulder with sweet potatoes, bake fruit cookies, and make raspberry jam (figs. 131–133).[23]

The Club Aluminum Utensil Company—an organization that sold its goods through house party demonstrations many years before Tupperware adopted the same strategy, thereby addressing housewives as consumers in their own homes—published a recipe book in 1926 featuring a vast array of succulent dishes.[24] Among its wares were a few items destined not for the kitchen but for the dining table, including electric doughnut and waffle molds (fig. 134). These small, seemingly insignificant objects were revolutionary inasmuch as they were among the first aluminum pieces to venture out of the rational sphere of the kitchen, where utility reigned supreme, into other living spaces of the home invested with complex cultural meanings. The items had stenciled patterns cut into their bases to give them a decorative appeal and render them appropriate within the new domestic setting.

The aestheticization of the aluminum object represented a major turning point in its early history. Until the mid-1920s, the appearance of aluminum cookware items had been completely determined by the technological and economic constraints of their manufacture and their utilitarian nature. Now, with the shift of the object from the kitchen into the dining area, and its growing importance as a symbol of feminine modernity, thought had to be given to its decorative appeal and meaning. There were essentially two solutions to this "design" problem in the 1920s: making aluminum objects traditional and "American," or imitating the look of other fashionable decorative objects of the day. In 1924, Griswold introduced a teakettle with a faceted body of stamped indentations reminiscent of silver and pewter decorative objects produced by leading French and Viennese designers of the period, among them Jean Puiforcat and Josef Hoffmann. The ad copy for this striking object— "Women just naturally like anything so rich in appearance as this superbly beautiful tea kettle"—made it abundantly clear that the kettle was meant to have "eye-appeal" and as such intended to seduce the female consumer who, it was believed, was always attracted to beautiful things.[25] In the same year, the Buckeye Aluminum Company advertised a similarly faceted teapot and accessories, this time with the object displayed on a dining table with a pastoral view depicted outside the window of a cozy room (fig. 135). The place of the aluminum object in the home was fast evolving, and its look was being modified to accommodate this further infiltration of the feminine sphere. As a result, aluminum objects could be seen as challenging the conventional decorative art objects currently at home in this arena.

To bridge the gap between tradition and modernity, Wagner launched its "Puritan teakettle" (fig. 136), a deliberate ploy to ally contemporary aluminum ware with the highly respected pewter ware of a century and a half earlier. The company's intention was to make its teakettle appeal to the housewife through links

## Mirro Roast of Beef and Browned Potatoes

A STANDING roast of the first two ribs is the best choice for a small family.

Score the fat side of the meat to prevent skin from shrinking and place the roast, bones down, on the perforated tray in MIRRO Roaster. Sear in a very hot oven (500° to 600°) for twenty minutes. Then season with salt and pepper, dredge lightly with flour and place cover on pan, with valve open. Reduce heat to moderate (about 300°) and allow 12 minutes to the pound for rare meat, 15 for medium and 20 if desired well done.

*BROWNED POTATOES.* Select potatoes of medium size and boil in jackets 10 or 12 minutes. Drain, cover with cold water, remove skins and place in roaster after meat is seared, turning them about to coat surface with fat. When meat is half done turn potatoes over and salt lightly. Serve potatoes around roast on hot platter.

Figure 131.
Cover of *Food Surprises from the Mirro Test Kitchen* (Manitowoc, Wis.: Aluminum Goods Manufacturing Company, 1925)
paper
cat. 3.91

Figure 132.
"Mirro Roast of Beef and Browned Potatoes," from *Food Surprises from the Mirro Test Kitchen* (Manitowoc, Wis.: Aluminum Goods Manufacturing Company, 1925), 3
paper
cat. 3.91

Figure 133.
"An Interesting Corner in the Mirro Test Kitchen," from *Food Surprises from the Mirro Test Kitchen* (Manitowoc, Wis.: Aluminum Goods Manufacturing Company, 1925), 17
paper
cat. 3.91

The Club Aluminum Equipment Complete

with American tradition, thereby imbuing it with a cultural value beyond that
of mere utility.

In the late 1920s, the aluminum object contained a fundamental contradiction
that was, nevertheless, an essential component of its meaning and acceptability.
Linked on the one hand with modern achievements in the areas of flight and warfare,
and with the idea of the housewife as a respected and efficient manager of her
home, and on the other hand with an aesthetic conservatism, it was a hybrid in
search of a unified identity. Not until the following decade, with the help of the
newly emergent industrial designer, did the unity of form and meaning become
an essential characteristic of the domestic aluminum object.

## Modernity and the Impact of Designer Culture

By the late 1920s, aluminum cookware had come a long way from the early years
of the century. It had achieved popular acceptance, become integrated into feminine
culture, and played a part in an ideological shift from domestic science to the more
emotional demands of motherhood and consumer culture. In the process, the
aluminum object had penetrated further and further into the everyday domestic
environment, and had been transformed from an essentially utilitarian item into

Figure 134.
"The Club Aluminum Equipment
Complete," from *The Recipe Book for
Club Aluminum Ware with Personal
Service* (Chicago: Club Aluminum
Utensil Company, 1926), 18–19
cat. 3.93

one with a degree of aesthetic appeal. Over the next ten years, all of these tendencies
would be pushed by the aluminum industry in another attempt to enhance sales.

But unlike the 1920s, the 1930s were dominated by economic depression and
an atmosphere of intense competitiveness among manufacturers vying to maintain
consumption levels. Many producers of new technology goods for the home and
office sought the skills of "visualizers" to make their products look more attractive
and to give them some "added value" in terms of "tastefulness." The collaborative
efforts of Raymond Loewy and the Gestetner Company, Walter Dorwin Teague
and Kodak, and Henry Dreyfuss and the Hoover Company are well documented.
As "designers for industry," these artistically gifted individuals, with professional
roots in advertising and retail, were committed to a modern aesthetic—called "streamlining," "streamform," or "streamlined moderne"—that took its inspiration from
both European modernist architecture and design, and the contemporary, dynamic
world of transportation.[26] Less well known, however, is the encounter between the
domestic aluminum object and the industrial designer within the context of modern
design culture in 1930s America, an encounter that would facilitate its final transformation from kitchen-bound utility item, to conservative decorative art object
in the dining room, to icon of feminine modernity in the domestic living area.

This shift coincided with, and arguably helped to create, yet another new
role for the housewife: that of hostess and upholder of the values of "casual living."
With the demise of the servant, and the increasing emotional and psychological
demands made upon the housewife, "casual living" and "informal hospitality" became
acceptable ways to juggle all the new demands of life in the domestic sphere. Once
again, promoters and salesmen of aluminum goods were quick to push the benefits
of their wares in the context of these changes but, for the first time, designers
came to the fore, keen not only to modernize the appearance of existing goods but

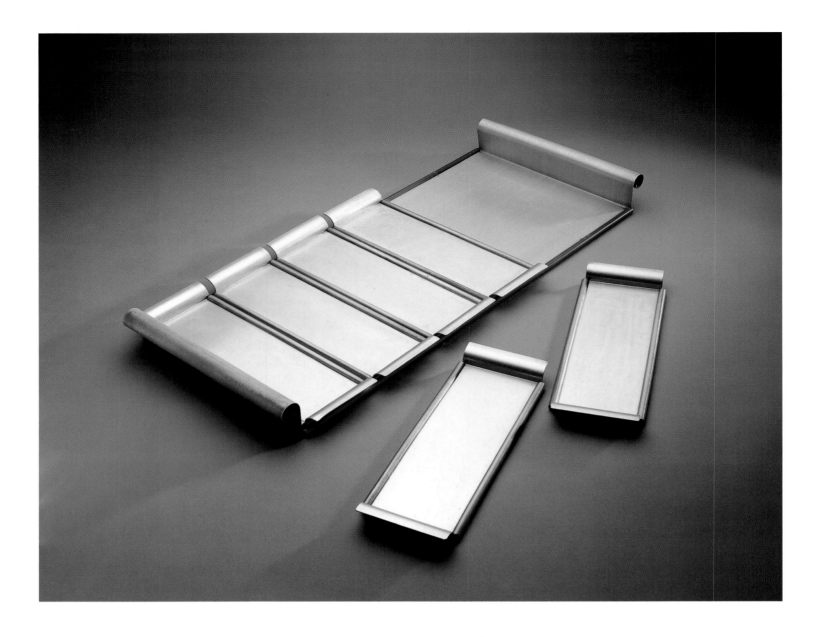

also to invent new objects and new functions for them and to invest them with new cultural meanings. Conventional items of kitchen-bound aluminum cookware, and the few aluminum items that had already ventured out of the kitchen (waffle molds and tea and coffee services, among them), were joined now by a wide range of new objects—serving pans, cheese boards, buffet trays, drinks trays, cocktail shakers, cigarette boxes, desk accessories, and candlesticks—geared to leisure and entertainment activities (figs. 137–138). The new aluminum goods quickly became desirable modern luxuries, destined for display as well as use, suggesting, outwardly, the social status and good taste of the owner, but inwardly, as the role of the house-wife changed beyond all recognition, becoming a sign of her self-worth.

The creation of new aluminum objects by designers enhanced their "cultural capital." Meanwhile, the utility aluminum object, in existence since early in the century, also assumed the mantle of high culture, moving from the sphere of banal domesticity into that of the "temples of art." It was not alone in this journey, as the modernist seal of approval for artifacts that spoke a unified language of function and form meant the elevation to high cultural status of a wide range of what had been seen as simple, utilitarian products, from radiators to automobile hubcaps.[27] These objects were transformed into "good design" and exhibited at shows such as *Machine Art*, held in New York's Museum of Modern Art in 1934. The Aluminum

Figure 137.
**Richard Neutra**
(American, 1892–1970)
Hors d'oeuvres tray, ca. 1930
anodized aluminum
cat. 3.38

Figure 138.
A. P. Company (American)
Cocktail shaker, 1934
enameled aluminum
cat. 4.9

Cooking Utensil Company displayed a wide range of its everyday kitchenware
in this exhibition, including some standard Wear-Ever items (stockpots, sauté pans,
a cake mold, a fruit press, and a teakettle). The functionalist tone of the exhibition
was confirmed in the catalogue with a quotation from St. Thomas Aquinas that was
said to have great relevance for the aluminum objects on display:

> For beauty three things are required. First, then, integrity or perfection: those
> things which are broken are bad for this very reason. And also a due proportion
> of harmony. And again, clarity: whence those things which have a shining colour
> are called beautiful.[28]

Two years earlier, an exhibition with a similarly functional agenda, entitled
*Design for the Machine*, was held at the Pennsylvania Museum (now the Philadelphia
Museum of Art). Aluminum objects were on display here as well, among them a
range of products created by the young designer Russel Wright, who was also responsible for the installation of an aluminum breakfast room. Wright began working
with aluminum in the early 1930s, discovering the craft technique of spinning the
material and finding that it was easier to work than pewter and chromium-plated
steel. His wife, Mary Einstein, seems to have played a part in persuading her husband

to design informal serving accessories and, a little later, oven-to-table ware, all made from spun aluminum but with wooden, cane, rattan, and cork components to soften them (and provide insulation) "so that their appearance suggests nothing of the kitchen."[29] It was she who elaborated new opportunities for offering hospitality in a casual way through conversations with magazine editors, radio interviews, and articles.[30] Einstein outlined a number of possible casual meals, among them the "Midnight Snack," the "Sunday Night Supper," the "After Theater Party," and the "Tin Wedding Surprise Party."[31] Wright designed and manufactured two series of aluminum utensils intended to enable the housewife to both create quick buffet meals and to be a relaxed hostess. They included a vast number of items, among them cheese servers, ice buckets, a bean pot, a coffeepot, a pretzel bowl, a tidbit stand, a canapé ball tray, a cocktail shaker, a mint julep tray, a beer set, and a relish "rosette" (fig. 139; Highlight 18). While Einstein defined consumer needs, Wright created the functions and forms of the objects designed to meet them. Feminine culture took the lead in this process, therefore, design following closely on its heels. Unlike the dining-room aluminum of the preceding decade, however, Wright's designs were highly innovative and unequivocally modern. They recalled pewter objects to a certain extent but did not copy their forms, nor did they mirror the styles of contemporary, European-influenced decorative arts. Rather, they openly embraced modernity, suggesting, through their brushed metal surfaces, a link with craftsmanship but, at the same time, through the designer's preference for rounded forms and their unadorned surfaces, an alliance with the machine age and the world of mass production.

Wright established his own means of manufacturing his designs at the beginning of the decade, and sales rose dramatically until he sold the spinning operation in 1939. He had played a key role in transforming the aluminum object into an icon of modernity, an accepted domestic decorative object that was closely integrated with feminine culture but, at the same time, manifested an uncompromising modern aesthetic. His spherical bun warmer, for example, with its wooden handle, harmonized with both the kitchen-based, streamlined domestic appliances and the simple blonde living-room furnishings designed by Gilbert Rhode and Wright himself later in the decade. In her novel *The Group*, which traces the lives of several Vassar graduates living in New York from 1933 onward, Mary McCarthy testifies to the role of Wright's objects as icons of luxurious modernity. As one of her protagonists exclaims: "Look at Russel Wright, whom everybody thought was quite the thing now; he was using industrial materials, like the wonderful new spun aluminum, to make all sorts of useful objects like cheese trays and water carafes."[32]

Wright's objects showed the aluminum cookware industry that "design" was an important factor in the marketing of their products, and that it was worth putting a designer on the payroll. Designers, the industry quickly realized, could fuse the beneficial qualities of aluminum with modern forms and modern lifestyles, thereby turning aluminum objects into powerful symbols of modernity as it was experienced by women in the domestic sphere (fig. 140). The lesson was learned very quickly. Wright's solitary achievement was quickly emulated by the mainstream industry,

Figure 139.
**Russel Wright,** designer
(American, 1904–1976)
Bun warmer and tidbit stand, ca. 1932
spun aluminum, rattan, reed,
and maple
cat. 3.74 and 3.75

Figure 140.
**Alfred Edward Burrage,** designer
(British, dates unknown)
Burrage & Boyd, now owned
by Staffordshire Holloware Ltd.,
manufacturer (British, 1932–present)
Picquot Ware K3 kettle, ca. 1950;
designed 1937, manufactured 1939–40
and 1948–present
polished die-cast aluminum-
magnesium alloy and sycamore
cat. 3.66

and by 1934, the *Aluminum News-Letter* announced the aluminum cookware
industry's new strategy:

1. Find out what people want
2. Commission a good designer to work out the problem practically
   and attractively
3. Produce it.[33]

Nowhere was this more evident than in the production of Alcoa's new line
of Kensington Ware, launched in 1934 by a new subsidiary, Kensington Inc. There
is no doubt that the decision to manufacture utensils for casual dining and other
display items, made from die-cast aluminum alloy, was a response to the success of
Russel Wright's aluminum goods. Alcoa's relationship with Lurelle Guild, the designer
who was commissioned to make it happen, dated from 1930. All industrial designers
actively pursued industry commissions, and Guild was no exception. A letter of 1930
from W. C. White of Alcoa addressed to Guild thanked the latter for a copy of his
book, *The Geography of American Antiques*, and ended: "I hope 1931 will bring you
everything you want—including an opportunity to work on aluminum."[34]

Clearly, Guild was keen to move into this area and by the middle of 1932
had worked on three products for Alcoa. As an article in the August issue of *Sales
Management* explained, "Since January a percolator, tea-kettle and pitcher have
undergone the redesign cure of Lurelle Guild, independent designer, in collaboration
with the production department of the Aluminum Cooking Utensil Company."[35]
The emphasis was clearly on "redesign" rather than product innovation, and Guild's
role was to simplify the objects in question and eliminate the swing handle of the

Figure 141.
(center) **Lurelle Guild**, designer
(American, 1898–1986)
Kensington Inc., manufacturer
(American, 1934–65)
Kensington Ware Stratford
compote, 1934
aluminum and plastic
cat. 3.85

(left and right) **Samuel C. Brickley**,
designer (American)
Kensington Inc., manufacturer
(American, 1934–65)
Pair of Kensington Ware Stratford
candlesticks, 1939
aluminum and plastic
cat. 3.86

kettle. With the formation of Kensington Inc., and the awareness of the need to compete with Russel Wright items, however, Guild was encouraged to be much more adventurous in designing a series of products that conformed to the idea of "casual living" but would also provide a modern equivalent to older "luxury" metals such as silver and pewter in a range of decorative items for the home.

Kensington Ware covered a wide spectrum of functions. By 1935, these included:

> table and service items for both formal and informal occasions; decorative
> pieces that sparkle with originality and have a most interesting variety of uses;
> drinking appurtenances having that look of tomorrow that is demanded by
> the discriminating lot; smoker's articles that are solidly practical and novel in
> competition; and desk accessories which happily wed the functional with
> the decorative.[36]

Stylistically, Kensington Ware ranged from a simple contemporary look to a piece inspired by a classical example to others inspired by the "Chinese Lowestoft pattern."[37] Crystal and brass were used to offset the visual effect of the aluminum (fig. 141). The names of individual items—the Laurel vase, the Zodiac platter, the Sherwood vase, the Mayfair coffee server, the Virginian cigarette box, and the Charleston tobacco jar—gave clues to the meanings they were intended to convey, which combined tradition with sophistication. The phrase "elegance without extravagance," used extensively in Kensington Ware advertising, completed a picture of tastefulness combined with a modern sensibility and definition of luxury.[38]

From the outset, Kensington Ware items were conceived as gifts, and were generally sold in giftware rather than house furnishings departments.[39] However, they also found a place in the silver department, where they soon were seen as modern alternatives to goods made from that precious metal. They were aimed at a well-off, taste-conscious market, and surveys of store sales showed that they were popular as gifts, especially for weddings. A line of overtly masculine goods was developed, for instance, a desk set and military brushes, promoted as ideal gifts from a woman to a man.

The acceptance of Kensington Ware was not achieved without huge promotional efforts by the manufacturer. Advertising was extensive in national women's magazines, and much effort was dedicated to the development of shop window and in-store displays, most of them designed by Guild. In 1934, for example, a permanent display room was opened on the twenty-first floor of the RCA Building at Radio City, New York (Highlight 19). Guild's design for that space was strikingly modern and lent an air of sophistication to the goods on display. Indirect lighting was used imaginatively and the color scheme was carefully controlled. Kensington Inc. took a strong hand in all marketing matters and worked closely with Guild to ensure that the company's identity and the brand image of its goods were given rigorous attention at all times.

By the late 1930s, it was clear that Kensington Ware had proved a profitable venture for Alcoa, and that working with a consultant designer had been a successful strategy. More than this, however, its existence confirmed that aluminum ware had moved completely out of the kitchen and become fully integrated into the dining room and leisure areas of the home. The shift from aluminum object as artifact necessary on rational grounds to one linked to the nonrational areas of life—beauty, desire, display, consumption, leisure, status, modernity, individual identity—was complete. It was a shift that paralleled the changing identity of the modern housewife as she was transformed from scientist to consumer.[40] By consciously following customers' aspirations and ideals, and shaping its promotional campaigns (and eventually its products) to fit changing ideals of femininity between 1900 and 1939 —indeed, helping to create these ideals—the aluminum industry had remained buoyant and financially strong.

Figure 142.
Reynolds Metals Company
advertisement
*Fortune* (April 1946): 8

### Conclusion

The first four decades of this century saw a dramatic shift in women's lives and value systems. The role of the aluminum industry in this transformation cannot be underestimated. Through its need to create a mass market, it participated in the construction of a consumer culture, and in the growing preeminence of women within that culture. It was a complex transformation, developed in stages and made possible by the work of a wide range of people in the commercial world of goods manufacture and promotion, from marketing men to advertisers to salesmen to retailers to designers. The housewife-consumer also played a role by confirming the place of the aluminum object in the home by the end of the period. The object itself, transformed from a utility item to an artistic object of desire, was not an innocent bystander (fig. 142).

During World War II, aluminum was drafted for military purposes. Consequently, manufacture of the decorative aluminum object declined and it never fully recovered its place within modern consumer culture, as it was upstaged—literally and symbolically—by other, newer materials. Its utilitarian role in the kitchen and its place in casual entertaining were challenged by the advent of stainless steel, ovenproof glass, and plastics. The fear that, as a cooking utensil, aluminum contributed to the spread of disease resurfaced with a vengeance in the years after 1945. As a modern luxury material, used for display and household items, stainless steel usurped aluminum's position to a significant extent. Within the context of aluminum's diminishing appeal as a consumer material, manufacturers turned their focus to the application of its properties in areas for which it was best suited—the internal components of electrical goods, air transportation, and so forth. The era of the light silvery metal being an answer to every housewife's dream had come to an end.

## Notes

1. Ruth Schwartz Cowan, *More Work for Mother: The Ironies of Household Technology from the Open Hearth to the Microwave* (New York: Basic Books, 1983), and Judy Wajcman, *Feminism Confronts Technology* (Cambridge: Polity Press, 1991).

2. The term "Mrs. Consumer" is taken from the title of the book by Christine Frederick, *Selling Mrs. Consumer* (New York: Business Bourse, 1929).

3. Information taken from "Early Product Letters," Folder 26, Series 1x, Box 1, Archives MSS*282, Aluminum Company of America Archive, Historical Society of Western Pennsylvania, Pittsburgh.

4. See correspondence between Griswold and Pittsburgh Reduction Company of 1896 and 1897 in Folder 11, Aluminum Company of America Archive, Historical Society of Western Pennsylvania, Pittsburgh.

5. See "Trade Notes," *The Aluminum World* 1 (November 1894): 36.

6. Ibid.

7. Folder 11, Aluminum Company of America Archive.

8. "Silver & Co.'s New Catalogue," *The Aluminum World* 1 (March 1895): 108.

9. Christine Frederick, *The New Housekeeping: Efficiency Studies in Home Management* (Garden City, N.Y.: Doubleday, Page, 1913).

10. Earl Lifshey, *The Housewares Story: A History of the American Housewares Industry* (Chicago: National Housewares Manufacturers Association, 1973), 166.

11. *The Wear-Ever Kitchen* (New Kensington, Pa.: Aluminum Cooking Utensil Company, ca. 1912), 4.

12. Ibid., 14.

13. *House Furnishing Review* 29, no. 4 (October 1908): 31–34.

14. *Aluminum Facts* (New Kensington, Pa.: Aluminum Cooking Utensil Company, 1906).

15. E. J. Presby, *Instructions to Salesmen* (New Kensington, Pa.: Aluminum Cooking Utensil Company, 1907), 25.

16. "Snappy Advertising Talk," *House Furnishing Review* 28, no. 2 (February 1908): 43.

17. "Equipping the Kitchen," *House & Garden* 37, no. 2 (February 1920): 52.

18. *House & Garden* 37, no. 4 (April 1920): 15.

19. *House & Garden* 36, no. 3 (September 1919): 61; 36, no. 4 (October 1919): 67; 36, no. 6 (December 1919): 5.

20. "To Better Appreciate Modern Housewares," *House Furnishing Review* 52, no. 2 (February 1920): 133.

21. "Housewares: The Changes Ten Years Have Brought About in This Fast Growing Industry," *House Furnishing Review* 52, no. 2 (February 1920): 142.

22. *The "Wear-Ever" Specialty Equipment* (New Kensington, Pa.: Aluminum Cooking Utensil Company, n.d.).

23. *Food Surprises from the Mirro Test Kitchen* (Manitowoc, Wis.: Aluminum Goods Manufacturing Company, 1925).

24. *The Recipe Book for Club Aluminum Ware with Personal Service* (Chicago: Club Aluminum Utensil Company, 1926).

25. *House Furnishing Review* 60, no. 3 (March 1924): 10.

26. See Penny Sparke, "'From a Lipstick to a Steamship': The Growth of the American Design Profession," in *Design History: Fad or Function?* (London: Design Council, 1978); Donald J. Bush, *The Streamlined Decade* (New York: George Braziller, 1975).

27. Philip Johnson, *Machine Art* (New York: Museum of Modern Art, 1934).

28. Ibid., 8.

29. *Russel Wright: Informal Tableware* (Detroit, 1925), 3.

30. *The New American Etiquette of Informal Entertaining (Proposal Suggestion for Radio Talk)*, Box 38, Russel Wright Archive, Syracuse University.

31. *Russel Wright: Informal Tableware*, 2.

32. Mary McCarthy, *The Group* (Harmondsworth: Penguin, 1966), 14.

33. "Juice Streams from This Stream-lined Juicer," *Aluminum News-Letter* (August 1934): 6.

34. Letter, Box no. 2, Lurelle Guild Archive, Syracuse University.

35. "Designing to Sell," *Sales Management* (August 15, 1932): 2.

36. *The Story of Kensington* (1955), Box no. 10, Lurelle Guild Archive, Syracuse University.

37. See "Tableware Goes Zodiac," *Modern Home* 7, no. 1 (April 1935): 2.

38. *Point of Sales Promotional Program for Kensington*, Box no. 10, Lurelle Guild Archive, Syracuse University.

39. As explained by the writer of a marketing exercise held in Kaufmann's department store in Pittsburgh in October 1934, "The Kensington pieces were taken from the Housefurnishing floor on Thursday and previous to this it seemed useless to spend much time there because the line seemed to be out of place on that floor." Memorandum from C. G. Towne, Box no. 2, Lurelle Guild Archive, Syracuse University.

40. See Annegret S. Ogden, *The Great American Housewife: From Helpmate to Wage Earner, 1776–1986* (Westport, Conn.: Greenwood Press, 1986), chap. 5.

## Aluminum: A Competitive Material of Choice in the Design of New Products, 1950 to the Present

Craig Vogel

By the mid-20th century, aluminum had passed through three different identities. Initially, aluminum was a precious metal and one of the most expensive materials to produce. By the late 19th and early 20th centuries, however, aluminum had become a cheap raw material competing for low-end production applications. After World War II, aluminum became a high-performance, mass-manufactured material, its low cost and excellent weight-to-strength ratios making it one of the most efficient materials in the world. Moreover, aluminum was and continues to be an extremely effective material to recycle. Unlike plastic, it does not deteriorate during processing and thus can be recycled repeatedly. These positive attributes of aluminum were perfectly suited to the postwar growth of mass manufacturing and consumption.

During the 1950s, '60s, and '70s, aluminum became the material that best exemplified the attempt by industry and the United States government to apply the advances of science and technology to the design and fabrication of objects that would be accessible and affordable to the ever-expanding middle class in their every-day lives. When it became clear that the Allies would win the war, President Franklin D. Roosevelt called on American industry to create the jobs necessary to stimulate the economy after the end of hostilities. In an issue of Alcoa's *Aluminum News-Letter* in 1943, it was observed that aluminum was only one of several materials that could play a role in this conversion to a peacetime economy.

> It would be the wildest fancy to dream that wartime developments of any one material could solve the postwar economic problems—much as we would like to picture our stronger, more versatile, more plentiful Alcoa Aluminum Alloys in the part.... In our imagineering, you will certainly come face to face with the greater possibilities in Alcoa Aluminum. They alone are not enough to create the payrolls to keep fifty-five million workers in their buying roles.[1]

Aluminum manufacturers were positioned to supply vast quantities of the material after the war. During the next five decades, the use of aluminum would help to transform modern transportation, sports and recreation, the storage and

Figure 143.
Heller Hostess-ware
(American, 1946–ca. 1955)
Colorama tumblers, sherbet
dishes, and pitcher, ca. 1950
anodized aluminum
cat. 4.49

consumption of food, even the way we sit and stand. Aluminum has also made it possible for many people to remain somewhat mobile and independent regardless of age and in the face of disabilities.

In the postwar period, aluminum played a major transitional role between the materials used in the early stages of the Industrial Revolution and the emergence in recent times of high-performance space-age materials. While aluminum would supplant copper, brass, iron, steel, and glass in many applications, in turn kevlar, titanium, boron, and carbon fiber would challenge aluminum's preeminence. Aluminum often made accessible to the middle class of developed nations a greater range of goods at reasonable prices. Manufacturers used aluminum to reduce the cost and weight of products, while at the same time increasing performance. Aluminum allowed producers and consumers to do more with less.

Aluminum suppliers continued to develop their manufacturing standards and capabilities during this period, and each decade saw new applications emerge. Several key products exemplify the use of aluminum, both alone and in combination with other materials, in ways that have had an impact on everyday life and reflect the concerns of consumers and society of their time. For example, the Colorama anodized tumbler and the tubular aluminum folding chair represented the optimism of the 1950s, while the designs of Charles and Ray Eames for the Aluminum Group and the Tandem Sling Seating System expressed the growing corporate and consumer sophistication of the 1960s. The aluminum beverage can helped beer and soft-drink companies respond to the environmental movement in the late 1960s and early '70s. Aluminum tennis rackets and baseball bats represented the growing interest in technology-enhanced sports performance in the 1970s and '80s. Finally, the aluminum walker is now a symbol of the graying of America and the ambition of elders to remain active throughout their lives.

The design of these aluminum products, as described in the following pages, has been affected by wider trends in American culture, and the products in turn have affected culture through the positive and negative impacts of their use. Their creation reflects and tells us much about two basic and distinct types of design: one driven by engineering, the other by industrial design. Engineers are more concerned with material performance and manufacturing; industrial designers are more sensitive to ergonomics and aesthetics. Integrating both types of design results in the best products. The everyday objects chosen for discussion here demonstrate a range of processes and approaches to design using aluminum as one of the primary materials.

## Designing with Aluminum

Two types of seating furniture—the tubular aluminum folding chair and the Tandem Sling Seating System—represent extremes in the design of furniture for high-use applications. The tubular folding chair was based on an anonymous design shaped primarily by cost and manufacturing concerns. The Tandem Sling Seating System, on the other hand, was created by two of the most famous industrial designers of the 20th century, Charles and Ray Eames. Their seating system, produced by Herman Miller, exemplifies the innovative approach this husband-and-wife team brought to all their furniture designs.

The tubular folding chair is one of the most popular postwar aluminum products (fig. 145). As such, it was an integral part of suburban expansion in the 1950s. As suburban housing developments sprang up almost overnight, the aluminum folding chair became the seating of choice for the backyard patio. It had many positive attributes: it was affordable, light-weight, did not rust, and was easy to fold, move, and store. Its plaid nylon woven seat and back complemented the plaid Bermuda shorts of the period, and soon the chair was a fixture around the barbecues and above-ground swimming pools that became integral to the suburban backyard summer experience.

The tubular chair had its origins in the aluminum tubing used as structural framing in aircraft during the war.[2] Alcoa called the process of finding new product applications for processes developed during the war "imagineering." After the decline in aircraft production, Alcoa and other suppliers worked with companies to develop new products incorporating aluminum. Their production capacity was matched in part by the need and desire of postwar families to equip their new homes with such products as the tubular chair.

The first folding chairs were an instant success. Everyone had to have them. The 1950s and '60s defined an era of mass consumption, especially among suburbanites. Women became the primary purchasers of home products, and these chairs were easy to maintain and light enough that any member of the family could move them. A 1954 advertisement for the Totalum Chair clearly directed its sales pitch to women: "There's no longer any need for the male member of the family to flex his muscles when it comes time to set up the folding chairs for lawn parties, bridge parties or outdoor meals" (fig. 144). These chairs could also fold flat and be taken to the town picnic, the lake, the beachfront, or to Little League games. The design fit the multi-use, action-oriented suburban culture. During cold weather, the folding chair could be conveniently stored in the garage or basement. It was soon available everywhere and became the type of low-end, inexpensive product that was sold in low-margin, high-volume department stores.

The aluminum tubular folding chair was equally successful in urban settings. During my youth in Brooklyn, it was common when walking down the street to see adults, especially seniors, sitting on aluminum folding chairs in front of their apartment buildings. The chair was light enough to carry to the elevator and then to the front of the building. Reflectors made of cardboard and a thin layer of aluminum foil were also a common sight in late spring and summer. Held at the base of the neck and unfolded to shoulder width, the reflector allowed sun worshippers to get an even facial tan in the comfort of their own yards. During the summer months, beaches were covered with aluminum folding chairs; some bathers preferred the standard type, while others sat in the sawed-off version known as beach chairs.

Figure 144.
Frederic Arnold Company
advertisement for the Totalum Chair
*Aluminum News-Letter* (May 1954): 3

Figure 145.
American
Folding chair, late 1960s
aluminum and nylon
cat. 4.47

The aluminum folding chair is one of the most popular chairs ever manufactured. It is not, however, one of the best chairs ever designed. Its economic efficiency far outweighed its appearance (aesthetics) and level of comfort (human factors/ergonomics). While its low cost, light weight, and portability made it a big seller, it had several serious design flaws that made it uncomfortable, unstable, and difficult to repair. The material used for the seat and back was often striped in combinations of green, white, orange, blue, and red plastic. Weaving the strips produced a garish plaid that visually overwhelmed the minimal aluminum tubular frame, which itself was easily damaged. While its light weight was an advantage for mobility, it was not ideal for stability—the chair easily tipped to one side, and even a small gust of wind could blow it around the backyard. The connectors holding the chair together were the cheapest, off-the-shelf solution, and there was little attempt to match the details to the whole. In sum, the aluminum folding chair was a group of parts assembled to respond quickly and inexpensively to an opportunity in the marketplace.

The chair also had several ergonomic flaws. The aluminum frame was contoured neither in front nor in back. After sitting in this chair for an extended time, circulation to the lower legs was often cut off by the straight tube running across the front of the seat. Leaning back for long periods was made uncomfortable by the tube at the top of the back. Where the seat meets the back, a thin bar connects the two and allows the chair to fold. Although the location makes structural sense, the bar meets the base of the spine and is thus a poor ergonomic solution. The armrest is a flat surface made of stamped sheet aluminum, a good choice economically but, again, a poor solution ergonomically. The plastic seating material has sharp edges and curls in hot weather, and the weave can leave an impression on the back and thighs. Moreover, if the seating strip rips, the chair weave slowly unravels. When it can no longer support the weight of an adult, the chair is usually thrown out.

The aluminum folding chair was the right chair at the right time. Symbol of 1950s American suburban culture, it reflected a period in which cost and function were considered more important than aesthetics. Donald Norman, author of *The Design of Everyday Things*, argues that humans have learned to adapt to poor design. They often see themselves, rather than the object, as the problem. This is certainly the case with the tubular aluminum folding chair.[3]

In contrast, the airport seating designed by Charles and Ray Eames, known as Tandem Sling Seating, is perhaps the best example of the use of aluminum in seating for public use. While more tubular aluminum folding chairs have been sold than Tandem Sling Seating units, more people may have actually sat in the Eames chair. All of the problems associated with the tubular folding chair are elegantly resolved in this design. Using primarily the same materials—an aluminum frame and plastic seating—the Eameses created an elegant, subtle solution that fit the context of a modern airport (fig. 146).[4]

All the designs produced by the Eames studio were the result of a collaborative process that addressed the needs of all concerned: client, manufacturer, primary customers, and secondary "customers" who came into contact with the product throughout its life cycle. Cost, material performance, aesthetics, and ergonomics were

Figure 146.
**Charles Eames,** designer
(American, 1907–1978)
**Ray Eames,** designer
(American, 1912–1988)
Herman Miller, Inc., manufacturer
(American, 1923–present)
Tandem Sling Seating (at John Wayne
Airport, Orange County, California),
1993; designed 1962, manufactured
1962–present

Figures 147–152.
**Charles Eames,** designer
(American, 1907–1978)
**Ray Eames,** designer
(American, 1912–1988)
Herman Miller, Inc., manufacturer
(American, 1923–present)
Assembly of Tandem Sling Seating

factored into the design. Beginning in the 1940s, the Eameses developed a series of chair designs utilizing first bent plywood, then fiberglass, and finally, in the 1950s, incorporating aluminum. They took the same approach to all their designs: minimizing materials and designing a strong and comfortable seat with a simple support structure.

The approach to Tandem Sling Seating, introduced in 1962, is far more comprehensive than that taken in the design of the aluminum folding chair. The designers, architects (Eero Saarinen for Dulles Airport, Washington, D.C., and C. F. Murphy & Associates for O'Hare Airport, Chicago), and manufacturer (Herman Miller, Inc.) shared a common vision: they all wanted a piece of furniture that would be the quintessential expression of modern design and complement the design of both the airport terminal and that most modern form of transportation, the jet airliner.

Working within the constraints imposed on them, the Eameses accepted the challenge to design seating for a dynamic, multi-use, demanding public space. Public seating must be easy to assemble, clean, move, service, and disassemble. The Eames design incorporates four basic materials: aluminum, nylon, steel, and urethane. The materials are used to their best advantage: cast aluminum for the seating frame, nylon for the seating surfaces, and urethane for the armrest padding. A steel beam provides structural integrity and determines the number of seats in the module. No excess pieces clutter the design, and the connectors holding the system together are hidden from sight to prevent clothing snags and reduce tampering.

Although it looks like it was made of stainless steel and leather, the system is actually produced at a fraction of the cost, weight, and maintenance of these materials. The design's understated elegance, typical of all Eames products, does not add to the visual complexity of the environment. The chair system is easy to manufacture, and the smooth, organic shapes of the aluminum parts require very little postproduction finishing (figs. 147–152). It can be shipped as a simple kit of easily assembled parts, which makes delivery cost-effective. The system is easy to maintain: the seating surface is stain-resistant and rugged and can be cleaned with a cloth, and the continuous form of the aluminum frame does not have small crevices where dirt can collect. The seating is easy to move and clean under. If the seat is damaged, one panel can be easily replaced.

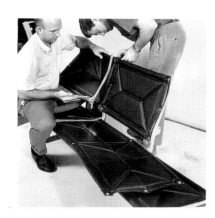  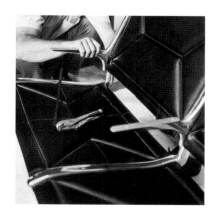

Moreover, the chair is comfortable. The back and seat are angled, creating a relaxed seating position (fig. 153). The armrests are well placed for relaxation and also provide support for getting in and out of the chair. The front of the seat is contoured to lessen pressure on the back of the legs. According to Arthur Pulos, author of *The American Design Adventure, 1940–1975*, the chair was designed to provide at least half an hour of seating comfort, the maximum time airlines expected people to wait. Pulos also notes that the armrest was designed to prevent travelers from sleeping horizontally on the seating system. Airline executives felt that travelers sleeping in waiting areas projected a negative image. The seat is wide and creates a private comfort zone that gives travelers a sense of space lacking in aircraft seats. It treats all potential passengers to the space normally available only to those traveling in first class.[5]

This is one of the few Eames designs that was widely accessible to the public. Many of the ideas developed in the Aluminum Group, designed several years earlier, were employed in Tandem Sling Seating. The Aluminum Group was intended to demonstrate that aluminum was a suitable material for elegant outdoor furniture. As a result of cost and limited distribution channels, the sales volume of the Aluminum Group did not come close to that of the aluminum folding chair, nor did it enjoy the widespread use of the Tandem Sling Chair. However, the Aluminum Group proved that aluminum could be used in outdoor seating that met the

Figure 153.
**Charles Eames,** designer
(American, 1907–1978)
**Ray Eames,** designer
(American, 1912–1988)
Herman Miller, Inc., manufacturer
(American, 1923–present)
Tandem Sling Seating unit loaded
on a pickup truck, 1965

concerns of manufacturing, ergonomics, and aesthetics. The success of Tandem Sling Seating was proof that the design concept could withstand significant abuse in a public space.

The differences represented by the approaches to the folding chair and the Eames chair still exist today. While it would be simple to refer to the Eames chair as more tasteful than the aluminum folding chair, the true difference lies in the approach to design. The aluminum folding chair was a direct translation of a military technology into a civilian application, a process commonly referred to today as tech transfer. The Eames design is a more thoroughly resolved chair, successfully balancing manufacturing methods, ergonomics, and cost. As consumer expectations rise, the comprehensive approach represented by Charles and Ray Eames continues to gain broader acceptance. Herman Miller, Knoll, and Steelcase are all making affordable furniture for home use, particularly for the home office. IKEA is an international company promoting good design at affordable prices, and even stores such as Target and Kmart are trying to improve their image through better design.

### The Anodized Rainbow

In 1950, Heller Hostess-ware (1946–ca. 1955) introduced a set of aluminum dishware called Colorama (fig. 154). The tumblers for this line were deep drawn and anodized in a rainbow of colors. Two decades later, Heller Designs, Inc. (1970–present) introduced a new line of dishware. This set, also produced in a rainbow of colors, was made of injection-molded plastic. Each line of dishware reflects the decade in which it was made.

The aluminum Colorama tumblers have recently reappeared as "collectibles" (fig. 143). Anyone who remembers using the original sets cannot help smiling when coming upon them for sale in a store or flea market. Aluminum tumblers, anodized in distinct metallic colors, are as strong a symbol of the naive optimism of the 1950s as the '55 Chevy. The tumblers and accompanying saucers and pitchers were a perfect complement to the tubular aluminum chair and the ideal modern accessories for the all-weather suburban picnic. Tumblers were even featured in advertisements for the Totalum Chair.

The tumblers certainly captured the spirit of the moment. The unique anodized metallic glow looked as though it might be a by-product of exposure to radiation, or be emitted by an extraterrestrial material. It was common for tourists visiting Las Vegas in the 1950s to take a picnic basket and watch the nuclear tests in the nearby desert (fortunately the wind tended to blow the radiation north of the city). The Colorama tumbler and the aluminum folding chair would have fit perfectly into this scenario. The anodized finish was referred to in Colorama advertisements as a "luster." Anodizing aluminum creates a finish that is part of the metal itself, not merely on its surface. It was unlike any material used in housewares at the time.

Aluminum's thermal conductivity precluded use of the tumblers for hot drinks. But they exhibited interesting qualities when used for cold beverages. The surface

# Exclusive NEW DESIGNS

Designed by BELLE KOGAN . . . I.D.I.

## DELUXE LINE
# COLORAMA® SERVING AIDS

**AMERICA'S MOST BEAUTIFUL ANODIZED ALUMINUM "HOSTESS-WARE"**

Colors fused into quality aluminum for permanent brilliance! PATENTED EXCLUSIVE ANODIZING PROCESS . . . won't fade, chip, tarnish or stain

**NEW** EXCLUSIVE DELUXE FOOTED TUMBLERS

*SMART 2-Tone COMBINATION*

*SMART 2-Tone COMBINATION*

**"COLORAMA" DELUXE 2½-QT. NEW ALUMINUM FOOTED PITCHER.** New, contemporary design in two-tone finish . . . beautiful Colorama. Satin finish aluminum foot and handle. Pouring spout fitted with ice guard. Colors*—Gold, Copper, Red, Green.
No. 401          Retail $5.00
(Please specify color.)

*Guaranteed by Good Housekeeping*

**"COLORAMA" NEW DELUXE 8-PC. FOOTED TUMBLER SET.** The most exquisite tumbler set ever produced! Heavy gauge! New Smart design in two-tone finish . . . beautiful Colorama color* combined with satin-finish aluminum foot. Set of 8 in assorted colors, 14-oz. size. In gift box.
No. 403 As above                                    Retail $7.00
No. 404 Set of 8, all 1 color. Colors: Gold, copper, red, green.   (Please specify color.)   Retail $7.00

**"COLORAMA" NEW, DELUXE SERVING TRAY.** Modern, oblong design with non-slip linear embossing. Size: 12" x 21½". Smart protective gallery and handles. Colors: Gold, Copper, Red, Green.
No. 402          (Please specify color.) Retail $6.00

*SMART 2-Tone COMBINATION*

**"COLORAMA" NEW DELUXE 7-PC. BEVERAGE SET.** New, contemporary design! 6 deluxe, heavy-gauge aluminum 14-oz. footed Tumblers in two-tone finish — Colorama color* combined with satin-finish aluminum foot. 2½-qt. Pitcher with ice guard in same two-tone combination, and with satin-finish aluminum handle. Colors*—Gold, copper, red Green. (Please specify color.)
No. 407          Retail $10.00

**"COLORAMA" LAZY SUSAN.** The tray of 100 uses. Brilliant, color.* anodized 14" Tray. 4 sectional, heat-proof milk-white glass dishes and covered casserole. Ball-bearing anodized-color* base and cover. Colors: Gold, Copper, Red, Green. Decorative black rings
No. 417          (Please specify color.)          Retail $7.00

NATIONALLY ADVERTISED

LIFE POST

GOOD HOUSEKEEPING

AND ON NATIONAL TELEVISION PROGRAMS

T.V.

NATIONALLY PROMOTED

**"COLORAMA" STEIN SET.** 12-oz. steins in assorted Colorama colors.* Ideal for iced drinks, small potted plants, etc. In attractive display box.
No. 6210—8-piece set, assorted colors          Retail $8.00
No. 6010—4-piece set, anodized copper only          Retail $4.00

**"COLORAMA" 8-PC. SHERBET SET.** A very colorful service for seafood cocktails, fruits, sherbets, desserts, etc. Each base finished in a different brilliant Colorama color.* Small crystal glass inserts.
No. 6347     As above          Retail $6.50
No. 6348     With gold bases and green inserts          Retail $6.50

*Anodized for lifetime brilliance . . . never needs polishing!

HELLER HOSTESS-WARE, WHITE PLAINS, N. Y.

had a distinct taste and texture. The aluminum took on the temperature of its contents and conveyed the coolness of the drink before it was even sipped. Condensation formed on the surface and kids drew on it with their fingers. Exposed on a kitchen shelf or set on a table, the tumblers reflected confidence in the postwar economy and represented another miracle product of imagineering.

Heller Hostess-ware began working with Alcoa in 1949 to develop manufacturing facilities for Colorama products.[6] Although they may have looked "neat," "cool," and "way out," the aluminum tumblers were appropriately named because they were inherently unstable. The form was deep drawn from sheet material and the manufacturing process determined its shape. It was wider at the top than the bottom, and extremely light. The tumbler had a uniform wall and base thickness, resulting in a high center of gravity, and thus, filled to the brim, would spill at the slightest touch. In fact, the tumbler was not ideal for barbecues or at home with young children actively reaching across the kitchen table. Although a new version was designed with an added base, it lacked the simple, direct shape of the original.

In use, the tumbler's perfect shape was quickly distorted as the soft aluminum bumped against other objects in the sink or dishwasher. The tumbler was easily scratched and its lip easily bent. Soon, it looked like a dented car body. As is the case with most modern designs composed of simple, smooth forms, any imperfection significantly detracted from the whole. The optimism and joy felt when purchasing these wares were soon diminished by their poor performance and rapid deterioration. The anodized rainbow of the 1950s gave way to the harvest gold and avocado green of the '60s. These later colors were achieved using an applied enamel finish.

In 1970, Alan Heller (son of the founder of Heller Hostess-ware) launched a new housewares company, Heller Designs, Inc. In 1972, he hired Massimo Vignelli to design a line of dishware.[7] Although similar in concept to the Colorama series— inexpensive, colorful, and casual—the Heller stacking ware was made of injection-molded plastic rather than aluminum. Following the trend of the 1970s—the last years of the International Style—to design products in perfect geometric shapes, Vignelli created a product based on the pure geometry of circles (fig. 155). Produced in a range of bright colors, the mugs formed perfect cylinders with no draft angle; their handles were shaped in a hollow half-circle. Plates and bowls came in graduated sizes, all circles with a simple plastic lip. When the multicolor set was stacked, it created a rainbowlike effect.

The Heller stacking ware had problems similar to those of the Colorama series designed two decades earlier. The set looked perfect when first purchased. Although plastics manufacturers had made great strides in improving the range and intensity of colors, the plates were easily scratched. The mugs crazed after repeated use with hot liquids, and the bottoms eventually cracked.

Both father and son chose to use a material in a new way. In 1950, aluminum was emerging from the pre-war perception as a cheap substitute for other materials. During the 1970s, plastic

Figure 154.
Heller Hostess-ware
(American, 1946–ca. 1955)
Advertisement for Colorama
Deluxe Line, ca. 1950

Figure 155.
**Massimo Vignelli,** designer
(Italian, b. 1931)
Heller Designs, Inc., manufacturer
(American, 1970–present)
Heller Dinnerware Stacking
Mugs, designed 1972,
manufactured 1972–present
plastic
cat. 4.48

was making the same transition. These two designs clearly demonstrate how quickly materials and colors can go in and out of fashion. It is interesting to note as well how the material determines the product's form and color. Both designs were produced in sets of rainbow colors, but anodized aluminum colors look completely different than the same colors in plastic. The failure of both to endure in the marketplace ties them forever to the era in which they were designed: seeing either product can immediately transport one back to a specific period in time.

While the Colorama series proved a less than optimal solution for a housewares product, it set the stage for one of the most effective uses of aluminum in the 20th century: the aluminum can. Aluminum manufacturers perfected the process of deep-drawing aluminum to form containers. During the 1960s and '70s, the aluminum can became the primary container for soft drinks and beer. The disadvantages that eventually caused the demise of the Colorama product line became positive factors in the case of the single-use aluminum can.

## Material Wars: Competitive Imagineering in Glass, Aluminum, and PET Plastic

The aluminum can, when combined with the aluminum pop-top opening system, is one of the best examples of material research and development in the 20th century. From the perspective of brand identity, the Coca-Cola bottle designed in 1915 is often considered the optimum package design. But from a performance standpoint, the aluminum can with its push-through tab opening is arguably the most successful container. While the "cola wars" between Coke and Pepsi are legendary, a largely invisible battle within the larger war has been waged among the materials manufacturers competing for the rights to produce containers for the lucrative soft-drink industry.

This battle is exemplified by the competition for market share in the bottling of Coca-Cola. The evolution of the Coca-Cola beverage container is well documented. A quick search on the Internet yields a number of sites for collectors of Coke memorabilia, and Coke bottles and cans are two of the most avidly collected items. Coca-Cola currently uses three different materials—glass, plastic, and aluminum— as containers for its soft drinks. Consumers have their own preferences, and beverage companies such as Coca-Cola use these material options in various ways to achieve their own marketing goals.[8]

By 1920, Coca-Cola had become one of the dominant manufacturers of soft drinks in the United States. This was accomplished by establishing a consistent identity centered on the 6.5-ounce Coca-Cola bottle, a strong advertising campaign, and Dr. John Pemberton's secret cola syrup formula. The Coca-Cola bottle is an icon of 20th-century American pop culture. The original bottle's organic shape made it easy to recognize and easy to hold.

During World War II, Coca-Cola established bottling plants, built at government expense, around the world. To help maintain troop morale, Coke followed the GIs wherever they went. After the war, Coca-Cola was positioned to become a major

international company. The 6.5-ounce bottle of Coke, priced at a nickel, dominated the soft-drink market until 1955, when Pepsi produced a 10-ounce bottle. Meanwhile, aluminum manufacturers were perfecting aluminum can production, and the aluminum can soon became the chief rival to the glass bottle. Kaiser Aluminum introduced the first aluminum can in 1956. Coors Beer was the first to use 7-ounce aluminum cans, in 1958. In 1965, RC Cola became the first soft-drink company to use 12-ounce aluminum cans. With so much equity in the glass bottle, Coca-Cola resisted aluminum packaging until 1967. Pepsi began using it in the same year.

Although tinned steel cans were used for beer and soft drinks from the late 1950s through the '60s, their weight, material cost, and the difficulty in opening their thick tops made them less than ideal competitors. With the refinements in aluminum can manufacturing, and the invention of the pull-tab opening, tinned steel became obsolete. The benefits of aluminum cans were clear to the beverage companies (fig. 157):

> Putting soft drinks in cans was an obvious solution. Canned beer had been
> found to occupy 64 percent less warehouse space than the same quantity
> of bottled beer and the shipping weight was less than half as much.... Savings
> were redistributed to advertising and also passed on to customers to induce
> them to change from bottles to cans.[9]

Consumers, however, still preferred glass, and even today liquor and wine are bottled in glass because of its perceived higher quality and the persistent belief that aluminum alters the taste of its contents.

But a number of factors would change consumer preferences. By the end of the 1950s, food shopping patterns had changed dramatically. Neighborhood deliveries and daily shopping at local grocers gave way to shopping at supermarkets. People drove to grocery stores and shopped once a week. Consumers began to buy soft drinks by the case. Glass bottles had to be returned and reused, a major responsibility for soft-drink companies and an annoyance for customers. Glass bottles had to be inspected and sterilized, and had a limited number of reuse cycles. The long thin neck, thick base, and indentations, which all helped to create Coke's early identity, became liabilities.

The packaging of a six-pack of Coke was very inefficient. The weight of a case of glass bottles full of Coca-Cola was more than the average shopper cared to lift. Separate six-packs were cumbersome. The bottle was not space-efficient, either for placing on supermarket shelves or for stacking in displays. Shipping glass bottles was more expensive, decreasing truck fuel efficiency. They were also heavy and difficult for delivery men to handle. Glass was susceptible to breakage at any point in the process.

The glass bottle had other disadvantages. Its crimped metal cap, designed to discourage tampering, was difficult to remove without an opener and, if done incorrectly, could chip the mouth of the bottle. As consumers demanded larger sizes, the Coke bottle became even more problematic because it did not lend itself easily to a larger size.

When Coke tried to switch to nonreturnable glass bottles in the late 1960s, environmentalists quickly reacted and Coke withdrew the idea. A survey conducted

Figure 158.
Coca-Cola Company advertisement for six-packs of aluminum cans, 1967

Figure 159.
Coca-Cola cans with push-through and pull-tab openings, 2000

in the early 1970s found that 5 percent of the solid waste in the United States consisted of containers produced by Coca-Cola.[10] Aluminum manufacturers set out to gain a larger share of the soft-drink packaging industry by solving the problems presented by glass. In contrast to the glass bottle, the aluminum can is a highly efficient container. The weight of the can itself does not add significantly to the total product weight when filled. The aluminum can when filled is an extremely stable stacking container (fig.158). Cases of soft drinks containing twenty-four cans can be stacked safely and efficiently in trucks, storerooms, or store aisles. A six-pack is a more efficient container for storing on supermarket shelves. And aluminum is easily recycled, in any condition.

The aluminum surface accepts printing and does not require an extra label. Each can becomes a cylindrical billboard clearly identifying its brand. To maintain a distinct brand identity, Coca-Cola used an all red can with white type and a ribbon shape that ran the length of the can. This curve, referred to as the "dynamic ribbon device," was based on the silhouette of the original bottle. The 1999/2000 Coke can includes an illustration of the original bottle.

The invention of the pull tab (the first patent was issued to Ermal Fraze in 1963) eliminated the need for a can opener. The development of the pull tab was a major engineering accomplishment. Engineers had to create a sealed container that could be opened by hand with a very low incidence of failure. The fact that pure aluminum cannot be welded posed a major challenge. The solution—a pull tab attached to a pin on the lid—placed incredible demands on manufacturing tolerances. The scoring of the lid had to be thin enough to pull away, but strong enough that it wouldn't explode under pressure.

The first pull tab had a few flaws, resulting in an unacceptable failure rate. The handle could break away completely or remove only part of the aluminum top. Once opened, the pull tab, which separated completely from the can, was discarded, and it quickly became a nuisance. Its sharp edge was dangerous and it was littered everywhere. In 1973, the pull tab was replaced by a push-through tab, which remained connected to the can (fig.159).[11]

Just when it seemed that aluminum cans had become the containers of choice, a new competitor emerged: the plastic bottle. The PET (polyethylene terephthalate) plastic bottle was developed at DuPont in 1975. Initially used only for one- and two-liter bottles, it has made significant inroads into the individual serving container market.[12] Just as the 6.5-ounce glass bottle lost out to the 12-ounce aluminum can, the 20-ounce resealable plastic bottle has displaced other rivals (fig.156). Consumers now want larger soft-drink bottles with easy-to-open resealable caps. Although consumers still buy soft drinks by the case, they also purchase them individually from vending machines and convenience stores.

With the desire to maintain a strong brand identity, Coca-Cola attempted to recapture the look of its original glass bottle in PET. In 1994, the company introduced a 20-ounce plastic version of its famous contour bottle. The new plastic bottle instantly differentiated Coke from other soft drinks packaged in straight-walled bottles, forcing other beverage companies to respond. When the trend to bottled drinking water began, plastic bottles emerged as the dominant material of choice. Aluminum cans are not as competitive for bottling noncarbonated water because water does not sell well in an opaque container that has a history of affecting the flavor of its contents.

Although plastic has many of aluminum's advantages, plastic bottles are far from perfect. The seal of a plastic bottle cannot maintain internal pressure and carbonation over time nearly as effectively as the aluminum can. Its thin walls are so soft that holding on too tightly when opening the bottle can cause overflow. The 20-ounce plastic Coke bottle requires a plastic cap and separate printed label, and the bottle must be made of two types of PET plastic, one for the base and one for the main body. The base is thicker than the walls of the bottle and thus creates a more stable footing. This is a complex container to produce and required the same type of research and development by plastics manufacturers that aluminum companies invested in the aluminum pull-tab can.

Any advance by one materials manufacturer is usually met with a quick response from others. In response to the variety of shapes possible in PET bottles, aluminum manufacturers have developed a process for producing three-dimensional relief on cans. This is yet another example of what Alcoa meant by imagineering.

During the 1970s, aluminum cans and pull tabs were under attack by environmentalists. In his book *The Evolution of Useful Things*, Henry Petroski notes that in 1975, only one in four cans was recycled; by 1990, the rate had risen to 60 percent in the United States. Over a period of twenty years, aluminum manufacturers had developed a system that made cans one of the most efficient materials to recycle. Recycled cans are essential to supplement the general aluminum supply, and the collection infrastructure is now so efficient that the metal in a used can may be recycled into a new one in as little as six weeks' time.[13]

## Sports Technology: Speed, Power, and the Sweet Spot

Over time, aluminum has replaced traditional materials in various types of sports equipment. The wooden javelin and bamboo vault pole are but two examples of such equipment rendered obsolete by the new material. When aluminum replaced wood in skis and golf-club shafts, fans and official rule-makers were concerned that the use of aluminum would have too much impact on athletic performance. In tennis and baseball, the use of aluminum has significantly changed the way the game is played. The aluminum tennis racket clearly altered the game and was the first step in the continuing development of high-performance rackets. In baseball, aluminum bats have been so successful that rule changes have been made to reduce their impact on performance.

*From Art to High Performance: Wooden Rackets versus Aluminum Rockets*
Aluminum tennis rackets made their debut in 1967 at the U.S. Open. Only four players, including Billie Jean King and Clark Graebner, used aluminum rackets that year. The winner was John Newcombe, who still preferred to use a wooden

Figure 160.
(center) Wilson Sporting Goods
(American, 1916–present)
T2000 tennis racket, ca. 1975,
designed 1967
aluminum, nylon, and other materials
cat. 4.34

(right) Spalding Sports Worldwide,
Inc. (American, 1876–present)
Kro-bat tennis racket, 1960s
wood, nylon, and other materials
cat. 4.35

(left) Prince Manufacturing, Inc.
(American, 1970–present)
Oversize tennis racket, 1987; designed
1976, manufactured 1976–present
graphite, nylon, and leather
cat. 4.36

racket. The design for an aluminum racket was conceived by French sports equipment designer and former player René Lacoste. Wilson purchased the rights to manufacture Lacoste's design. Several other companies were soon producing variations on the Wilson T2000.

Seven years later, in 1974, Jimmy Connors defeated Ken Rosewall at Wimbledon and again at the U.S. Open. His victories, achieved using the Wilson T2000 aluminum racket, signaled the end of an era (fig. 160). His success became synonymous with the success of the new racket. Australians had dominated the game for the past two decades and, although a respected sport in the United States, tennis was not as popular as baseball, football, basketball, or golf. Jimmy Connors changed that with his success and the new style of play he brought to the game. When Chris Evert emerged as the queen of the women's game, she also chose to use the Wilson T2000 and the transition was complete.

The aluminum racket was one factor in the growth of amateur tennis and the increased interest in tennis as a spectator sport. The A.C. Nielsen Company conducted its first survey of tennis in 1970, estimating that 10.3 million Americans played occasionally. A third study released in 1974 indicated a staggering 68 percent jump to 33.9 million who said they played tennis "from time to time," and, more significantly, it was estimated that 24.3 million people played at least three times a month:

> Almost as surprising as the rate of the participation boom was a Louis
> Harris survey that indicated a substantial rise in tennis' popularity as a
> spectator sport.... This growth was reflected in the tennis industry, as
> new companies rushed in to offer a dizzying variety of equipment to the
> burgeoning market.[14]

Figure 161.
The Smasher string and grommet design
*Spalding 1971 Retail Catalogue* (Chicopee, Mass.: A. G. Spalding & Bros., 1971), 62–63

The aluminum racket made it easier to hit the ball. Its larger "sweet spot"[15] reduced torquing, which causes the handle to rotate in the player's hand, thereby sending the ball off target. The aluminum racket was more durable than the wooden one, which warped easily, had thick rims, smaller hitting surfaces, and smaller sweet spots. Aluminum frames were thinner and lighter, and would not warp if left out in the rain. Nor did they have to be stored in clamps between matches. Professional and amateur aluminum rackets were much closer in quality than low- and high-end wooden rackets. Amateurs could finally enjoy playing tennis, regardless of their talent.

The production of wooden rackets was as much an art as a craft. The manufacture of aluminum rackets was the consequence of developments in technology. This racket was designed for mass manufacturing. Its frame is extruded aluminum which, when bent into a racket shape, is stronger than when straight. The extrusion proved to be so structurally sound that manufacturers could increase the size of the racket head, creating even more forgiving oversized rackets for beginners.

In 1968, Peter A. Latham and Paul Brefka designed an aluminum racket for Paul Sullivan Sports, Inc. Reynolds Metals Company served as manufacturing consultant. The designers chose aluminum after researching a number of materials:

> The designers analyzed fiberglass, steel, wood, magnesium and aluminum and found that aluminum best met structural and manufacturing requirements. The state of the art of fiberglass was not yet sufficiently refined in 1960–70; graphite was still only utilized at the aerospace level and not economically viable; boron was still in the laboratory stage.[16]

With the advent of color television, more sophisticated camera coverage of the games, and the development of the tie-breaker, more people began to watch tennis on television. Professional leagues emerged and tennis was played year round. The popularity of tennis as a spectator sport fueled fan interest in actually playing the game. The increase in the number of composite courts, both indoors and out, made it easier and cheaper to play. Amateurs could sustain longer volleys playing on consistent surfaces with more forgiving rackets. All of these developments were part of a greater interest in staying fit.

Wooden rackets had ruled the game since its inception centuries ago. The aluminum racket, in turn, was quickly challenged by a variety of new material options. Carbon fiber, composites, and titanium are only three of the materials that have replaced aluminum at the professional level. Wilson is now selling a variety of T6000 and T7000 rackets featuring the latest flex technology and space-age materials.[17] The best players are now serving at speeds between 120 and 130 miles an hour. There are fewer long rallies in the men's game each year, as most top players have developed a serve-and-volley game. The new rackets are primarily responsible for the increased speed of the game. All of these current advances are based on innovations that began with the aluminum extruded tennis racket (fig. 161).

*Do You Prefer a Crack or a Ping?*

On August 12, 1998, the National Collegiate Athletic Association (NCAA) passed a ruling destined to reverse the technological advances made in the design of aluminum bats. The NCAA stated that, starting in 1999, aluminum bats would have to perform more like wooden bats. The ruling was made after the record-setting 1998 College World Series championship game, in which Southern California defeated Arizona State 21–14, and 52 records were tied or broken, including 35 runs scored, 39 hits, 34 RBIs, and a combined 8 home runs. Two factors were cited as reasons behind the decision to limit the aluminum bat's performance: the velocity of the ball after being hit by the bat was becoming faster than the reaction time of pitchers and infielders, and the increase in power-hitting was disproportionate to the natural capability of hitters.

Aluminum bats help batters swing faster, contain a larger sweet spot, and have greater flex in the handle (fig. 162). These qualities increase the likelihood that the bat will make solid contact with the ball. Aluminum bats were originally chosen for use in Little League, high school, and college games because they performed more consistently and were more cost-effective than easily broken wooden bats. The first aluminum bats, designed in the 1930s, could be used only for softball. They had clear advantages:

> The Carpenter aluminum bat is similar in weight and balance to the regulation wood softball bat, yet it has a much longer life and cannot be damaged if left out-of-doors… and never warps or splinters. It is the ideal bat for playgrounds and institutions where equipment is put to severe and careless use.[18]

Several decades passed before aluminum bats became serious competition for wood. The long-standing tradition of wood and the prejudice against aluminum made it difficult for athletes and coaches to see the value in an aluminum bat. During the past few years, competition among aluminum bat manufacturers has resulted in bats that are incredibly powerful and far more forgiving than wood; however, prices have risen and the life of the bats has decreased significantly. While aluminum bats do not break, they deform and flatten out after repeated use.

The performance of aluminum bats has finally crossed the boundary of what is considered appropriate technology-assisted athletic performance. Babe Ruth, Lou Gehrig, Joe DiMaggio, and Hank Aaron all established batting records using wooden bats, but their numbers would become insignificant if today's professionals were allowed to use aluminum bats. If Mark McGwire and Sammy Sosa, for example, had been using aluminum bats in 1998, they could easily have hit more than 100 home runs. Should technology be capped to allow human capacity to remain the primary determinant of athletic success? Should technological advances always be allowed to displace traditional ways of doing things? It seems that in tennis, aluminum rackets improved the game on a number of levels. It is not clear that this is the case in baseball.

Figure 162.
Hellerich & Bradsby Company, manufacturer (American, 1884–present)
Easton Sports, manufacturer (American, 1922–present)
Softball and baseball bats, 2000
aluminum, graphite, leather, and rubber
cat. 4.29 and 4.30

Figure 163.
Hellerich & Bradsby Company, manufacturer (American, 1884–present)
Louisville Slugger wooden bats, ca. 1949

Figure 164.
Roberto Clemente (1934–1972) at bat for the Pittsburgh Pirates, Three Rivers Stadium, Pittsburgh, September 30, 1972

The production of wooden bats is tied to the human history of woodworking, carrying with it centuries of tradition (fig. 163). Every wooden bat is unique, and makes a distinct cracking sound upon impact with the ball. A knowledgeable fan hears this sound and can tell how well the ball was hit. Every aluminum bat is an exact replica of the previous one. The aluminum bat makes one sound: a technological ping. New professional parks are being designed in retro styles, a trend begun with the building of Camden Yards for the Baltimore Orioles. Given this trend in stadium construction and the recent actions taken at the collegiate level, it is unlikely that professional baseball players will be allowed to use aluminum bats. In this case, it seems better to keep wooden bats at the professional level and to promote the use of aluminum bats that have performance characteristics closer to wood at other levels.

High-performance athletic shoes, the use of instant replay for assisting referees in football, and the sensor system used at Wimbledon to call serves are some of the many advances in sports technology that have generated controversy. During the past two centuries, change has occurred at a rate faster than human systems can adapt. There is a need to develop better methods for anticipating the positive and negative impacts of emerging technology, and the wisdom to make thoughtful choices in its integration.

## Aluminum for the Ages

Today, more than 31 million Americans are over the age of sixty-five, and their number is growing at twice the rate of the general population. The fastest-growing group comprises those over the age of eighty-five, among whom chronic illness and increased frailty are more evident.[19] These demographic trends are occurring in all industrial nations. The terms "universal design" and "transgenerational design" (attributed to Jim Pirkl) describe a new approach to product design that recognizes and accommodates these developments. Products, according to this approach, should be easy to use regardless of age or disability. The American Disabilities Act (ADA) has an even broader agenda, requiring that buildings, environments, and products be accessible to all people regardless of physical and cognitive ability.

An early example of universal design is the aluminum walker (fig. 165). The aluminum walker is helping people with limited ambulatory ability lead more active lives. The walker has become, at the beginning of the 21st century, what the aluminum folding chair was at the middle of the 20th. As parents of the baby-boom generation become senior citizens, many of them are using aluminum walkers as they recover from surgery or injury, or as a long-term aid in staying mobile. The need to remain active has generally overcome fear of stigma attached to using a "crutch" or being viewed as disabled. Independent mobility, even with the aid of a walker, is crucial to maintaining a sense of dignity and self-worth.

The design of the aluminum walker is perhaps the most significant application of aluminum since its use in aircraft during World War II. Functioning as a type

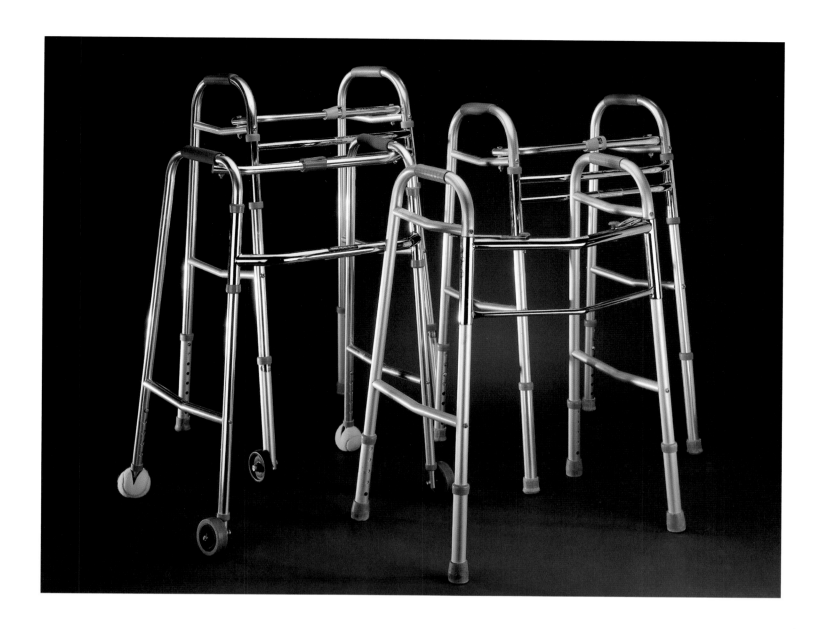

of exoskeleton, the aluminum walker has enabled millions of people to remain mobile who might otherwise have been bedridden or confined to a wheelchair. Techniques in rehabilitation therapy and surgery are allowing people of all ages to maintain more active lifestyles and to quickly become ambulatory following accidents, strokes, or surgery. The aluminum walker became an integral component of rehabilitation when health professionals realized that lack of activity can seriously hamper a patient's recovery.

The aluminum walker has a unique combination of structural integrity, light weight (which allows it to be lifted), and low cost. Although it is not the most elegant product, its simple, efficient, functional design is far more successful than the tubular aluminum chair discussed earlier. The basic walker can be folded and stored in a car trunk or closet. Although the walker requires the addition of a steel reinforcement bar and plastic and rubber details to allow for better grip, it is clear that aluminum is the primary factor in the success of this design. The product has worked so well that a number of alternative walkers are being marketed specifically for use in public spaces. Until the mid-1970s, people with limited mobility were rarely seen in public. Today, thanks to the walker, they are a common sight (fig. 166).

The aluminum walker is better than the alternatives, but there is room for improvement. The device looks quite clinical and elders often feel stigmatized when using one. Several new European designs are better-looking but more expensive,

Figure 165.
American
Walkers, 2000
aluminum and other materials

and are made primarily for outdoor use. There are no elegant, inexpensive solutions for walkers used indoors. Current versions can be awkward to negotiate—when sitting down to dinner, for instance—and create problems in senior care facilities when residents gather for entertainment or dining. An elevator's capacity is reduced when occupied by seniors with walkers, which cannot be used on stairs.

## Conclusion

Designing new products is a complex process involving a myriad of factors. Selecting the right materials and manufacturing methods are two important decisions in the product development process. Product developers must integrate their knowledge of materials and processes with an understanding of market forces operating at a given time. The best products match material with consumer preference, and not only connect with current trends but also resonate with deeper consumer needs for comfort, safety, and quality.

While all the products discussed in this essay were successful, some were more so than others, and in different ways. Compare the long-term success of the Tandem Sling Seating System and the aluminum beverage can to the relatively short-lived Colorama tumbler and aluminum tennis racket. Or the manufacturing success of the aluminum lawn chair to the social advances made by the aluminum walker in responding to the emerging issues of universal design. In the world of democratic capitalism and industrial growth, the process of product development is in a state of constant flux. The designer, consumer, and material manufacturer are part of a cycle that continually creates opportunities for new products. Aluminum will remain a competitive material of choice in new product development as long as manufacturers continue to develop innovations that respond to changing forces in the marketplace.

## Notes

1. "We Have This to Say About Lead," *Aluminum News-Letter* (October 1943): 8. This article is part of a series published in 1943 in Alcoa's monthly newsletter. Each article discussed the merits of each material and claimed that after the war there would be new applications for all materials.

2. Phil Patton, *Made in USA: The Secret History of the Things That Made America* (New York: Grove Weidenfeld, 1992), 99.

3. Donald A. Norman, *The Design of Everyday Things* (New York: Doubleday, 1989).

4. John Neuhart, Marilyn Neuhart, and Ray Eames, *Eames Design: The Work of the Office of Charles and Ray Eames* (New York: Abrams, 1989), 226–29 (Aluminum Group), 274–75 (Tandem Sling Seating).

5. Arthur Pulos, *The American Design Adventure, 1940–1975* (Cambridge, Mass.: MIT Press, 1988), 326.

6. "Heavenly Houseware!" *Aluminum News-Letter* (December 1952): 3.

7. Nada Westerman and Joan Wessel, *American Design Classics* (New York: Design Publication, Inc., 1985), 108–09.

8. Thomas Hine, *The Total Package: The Evolution and Secret Meanings of Boxes, Bottles, Cans, and Tubes* (Boston: Little, Brown, 1995), 160–63.

9. Ibid., 161.

10. Mark Pendergrast, *For God, Country, and Coca-Cola: The Unauthorized History of the Great American Soft Drink and the Company That Makes It* (New York: Macmillan, 1993).

11. Henry Petroski, *The Evolution of Useful Things* (New York: Knopf, 1992), 199–203.

12. Hine, *Total Package*, 161–62.

13. Petroski, *Evolution of Useful Things*, 199–203.

14. Bud Collins and Zander Hollander, eds., *Bud Collins' Tennis Encyclopedia* (Detroit: Visible Ink Press, 1997), 211.

15. The "sweet spot" is the area around the center of mass of a bat, racket, or head of a club that is the most effective part with which to hit a ball.

16. Wessel and Westerman, "Aluminum Tennis Rackets 1968," in *American Design Classics*, 144–45.

17. All the latest Wilson racket models are available for review at www.wilsonsports.com/Tennis/homepage.asp.

18. *Aluminum News-Letter* (February 1939): 6.

19. Bruce Tokars, *Aging Parents: The Family Survival Guide* (San Francisco: SyberVision Systems, 1996), 2.

Figure 166.
Man sitting on beach in folding chair with walker, ca. 1990

# Aluminum and the New Materialism

Paola Antonelli

In contemporary design, materials have a renewed status as catalysts of inspiration and as integral elements in the design process. During the technological boom of the 1960s, many new materials such as polyurethane foam, polypropylene, and medium-density fiberboard sparked entirely new typologies of objects. Not until the late 1990s do we once again see a design culture informed by either new materials or brand-new applications of traditional ones.

Aluminum belongs in this latter category. In the 1950s, aluminum was the material that expressed optimism about a bright and hygienic future. Today, its power resides in both its ability to evoke nostalgia for this past and the technical possibilities allowed by its flexible nature as a material (fig. 167).

Among the most effective and beautiful examples of aluminum's role in contemporary design are Hisanori Masuda's Iquom boxes (fig. 168). The pebblelike containers, made of sand-cast recycled aluminum and manufactured in Yamagata Prefecture in Japan, famous for its centuries-old tradition of metal arts and crafts, are a unique amalgam of old and new. Local material culture is merged with today's global material soul. Closed, the shiny boxes look like drops of quicksilver, but reveal a sanded heart of gold when opened. The recycled metal lends itself to unrestrained poetic gesture. These boxes are signs of our peculiar and sensitive times, symptoms of a sophisticated society that has learned to recognize patterns of beauty in pragmatic, economic ideas. And aluminum, economical and expressive, is the narrator. Contemporary design, in turn, is alive with experimentation and creativity, projected toward a sustainable future, optimistic and honest, aware and involved in social anthropology.

Our perspective on the material world has changed dramatically during the past ten years, so deeply as to visibly affect areas that are traditionally resistant to ethical tidal waves, such as fashion and popular culture. After the sensory and material overdrive of the 1980s, the jaded inhabitants of the West seemed ready for a new obsession, this time with simplicity and purity. The new code of action in the field of design is best illustrated by Droog Design (or "dry design"), the Dutch movement that celebrates the virtues of ingenuity and economy within a coherent minimalist aesthetic. The apparent modesty of Droog Design objects, made with recycled parts and apparently low technologies, has also made them emblematic of a new less-is-more approach worldwide.

Figure 167.
**Ross Lovegrove**, designer
(Welsh, b. 1958)
Bernhardt Design, manufacturer
(American, 1889–present)
Go chair, 2000; designed 1999,
manufactured 2000
380 aluminum alloy and
polycarbonate plastic
cat. 5.40

Droog Design made its first appearance in 1993, in an exhibition featuring a manifesto-like collection of objects by various designers, at a time when no one—neither design critics nor users—seemed able to tolerate visual and functional redundancy any longer. Droog's work ushered in what some labeled the "neominimalist era," a healthy and welcome systemic revolution that would lower everyone's blood pressure. Similarly, the fashion world began to celebrate the work of Miuccia Prada as well as Tom Ford's revival of the Gucci label. Prada and Ford took garments one step beyond the previous Japanese avant-garde with a peculiar blend of bare-boned yet high-tech simplicity. The exceptional Prada collections of those first years featured synthetic fibers and clear A-cuts reminiscent of hospital or stewardess uniforms—a hygienic, almost antiseptic twist on the concept of elegance. Metals, and aluminum in particular, are strongly suggested by not only the look but also the feel of many of the new synthetic fibers and finishes of choice.

In 1995, at the third meeting of Doors of Perception, a design conference organized in Amsterdam by the Netherlands Design Institute and that year entitled "On Matter," the trend forecaster Lidewij Edelkoort gave a poignant audiovisual presentation that ultimately made the audience feel clean and purified. She showed natural fibers, minimalist furniture, recyclable materials, and translucent bars of soap. The world, one could assume from her talk, wanted fewer, better, clearer, sounder things—and would be willing to pay more to have them.

In the same passionate search for cleanliness, Japan was the first country in which air was marketed as whiffs of pure canned oxygen offered for sale in vending machines, followed shortly by $O_2$ Spa Bar in Toronto, which offered scented oxygen inhaled through slim pipes inserted in the nostrils. In May 1997, at a symposium at the San Francisco Museum of Modern Art devoted to the theme "Icons," art historian Alexander Nemerov spoke eloquently about a new American icon: bottled noncarbonated water. He joined the chorus calling for a cleaner, purer lifestyle, and identified the concept that cosmetics company Neutrogena has been marketing

Figure 168.
**Hisanori Masuda,** designer
(Japanese, b. 1949)
Kikuchi Hojudo Inc., manufacturer
(Japanese, 1604–present)
Chushin-Kobo, manufacturer
(Japanese, 1997–present)
Iquom Tableware Collection "Egg"
and "Oval" jewelry boxes, 1999;
designed 1992, manufactured
1992–present
sand-cast recycled aluminum
and gold leaf
cat. 4.65

successfully for years: nothing is as deeply sensual as *nothing*—a transparent, odorless substance that cleanses the body inside and out.

We are awash, to borrow writer Kurt Andersen's words, in a cleansing cultural flood tide. Just as has fashion, the design landscape, too, has lightened up. During the past five years, furniture has become sleeker and less formally impertinent. Accessories have become smaller, more personalized, with tempered, pared-down interfaces. Interestingly, much of the identity of things today resides in surface treatments and color, which have made a cheerful return after a hiatus of many years filled by the black-and-chrome tyranny. This new phenomenon is assisted by the progress of technology itself, which allows for many more variations and degrees of freedom within the same manufacturing cycle—creating a much more interesting material world.

Dutch design of the mid-1990s also revealed the striking of a new balance between technology and artifacts that has enabled this systemic revolution. Contemporary design does not glorify advanced technology as it did in the 1980s but, rather, appreciates technology for its ability to simplify our visual and material landscape—in other words, for the way it can simplify our lives by making objects lighter, smaller, and less formally intrusive, as well as less taxing to the environment. Visual redundancy has been exiled to the Internet, which is still possessed by a chaotic and supranational design Esperanto.

Thus today, very high technology can coexist in peaceful synergy with very low technology. A cleaner lifestyle is the goal, and any means that will achieve this end—even craftsmanship—is deemed good. Besides, craftsmanship is no longer considered reactionary; advanced materials, such as the aramid fibers of Marcel Wanders's Knotted Chair, in which a rough knotted fishnet is frozen in shape by resin, and the fiberglass of Hella Jongerius's Knitted Lamps, in which the bulbs are covered by a translucent white stockinglike elastic fabric, can be customized and adapted by the designers themselves. On the other hand, the tools to work materials into a final shape might not yet exist. For this reason, some advanced materials require manual intervention, while some low-tech media, like recycled glass milk bottles, the use of which responds to ecological needs, demand a crafts approach because, as "found objects," they must be assembled by hand. Experimentation, be it high tech or low, requires a hands-on approach, and the flexibility and novelty of the materials and manufacturing methods available today have stimulated the exploration of numerous possibilities.

Contemporary design is therefore an interesting composite of high and low technologies. New technologies are being used to customize, extend, and modify the physical properties of materials and to invent new ones endowed with the power of change. New materials can be bent and transformed by engineers and by designers themselves to achieve their design goals. In a 1995 exhibition at the Museum of Modern Art in New York, I introduced the concept of *mutant materials* to illustrate the new design reality sparked by the progress in material technology and culture, according to which adherence to the "truth" of a material is no longer an absolute for design.[1] Materials have become curiously deceptive and sensitive to the designer's

intentions, like aluminum in Hisanori Masuda's boxes or in a screen by Fernando and Humberto Campana (fig. 169). Ceramic can look and act like metal, just as plastic can look like ceramic or glass, and wood can be soft. An object that appears to be made of metal and feels as smooth to the touch as plastic may instead be ceramic. Because wood veneer can now be applied directly to polyurethane foam, a hard and monastic-looking wooden stool can be surprisingly soft. A knotted rope can be as sturdy as metal. Plastic can be made to look as clear as glass, as sharp-edged as stone, and feel as metallic as aluminum. Scientists have found ways to rearrange the molecules of matter in new materials that not only appear different from those of the past but also exhibit distinctly new personalities and behaviors.

The misunderstood alchemists of the past must grin at the materialization of their dreams as the elements we once knew and recognized have become the basic ingredients for new, recombinant recipes, opening up a new world of possibilities for designers and manufacturers. No longer consigned to the passive role of adjunct, materials have been transformed into active interpreters of the goals of engineers and designers. The relationship between raw material and finished object was once linear and clear, and distinctions between objects made of the same material were inscribed by the techniques of manufacture. The mutability of a material, on the other hand, is not merely a function of the quantity and diversity of the objects that can be produced from it. Rather, it is related to the diversity of behaviors and personalities a material can assume *before* being shaped by the techniques of its application.

Many materials have found a new life in this fertile climate, and aluminum is certainly among them. While many primal media, such as wood and marble, can claim longevity and continuity across the centuries, few—glass and ceramic, for instance—can also claim timeliness and adaptability. Mutant materials par excellence, they have, at different times and places in history, been able to give shape and presence to diverse material cultures. They have morphed and adapted to become signs of the times. Aluminum, albeit much younger than glass and ceramic, has shown the same potential. Since its introduction in the 19th century, aluminum has signaled periods in design history as its uses have exploited, at various times, its preciousness, economy, ductility, strength, or lightness. While aluminum's history has been amply explored in this volume, a brief review is relevant here in light of the metal's contemporary use, which simultaneously takes advantage of all its multiple personalities.

At the time of aluminum's introduction in the second half of the 19th century, the earthquake of the Industrial Revolution had momentarily divided "good design" from manufacturing. The Arts and Crafts movement was raising issues of ethics and aesthetics in the manufacture of design, and its proponents, William Morris in particular, had drawn an ideal and somewhat bigoted pendulum for our future instruction. On one side of the pendulum, the righteous side, sat the craftsperson, independent maker of ideas and enlightened master of beauty and honesty. On the other side rested the industrial manufacturer, mere actuator of the evil and the ugly, setting materials against their own nature. Yet the best of design, history teaches us, is created in the intermediate stations. Aluminum—at that time precious,

Figure 169.
**Fernando Campana,** designer
(Brazilian, b. 1961)
**Humberto Campana,** designer
(Brazilian, b. 1953)
Campana Objetos Itda-ME,
manufacturer
(Brazilian, 1984–present)
*Sculpture Screen*, 1993
reused aluminum television
antennas and aluminum wire

new, and crafted—lent itself to artistic explorations in castings. It was on the "good" side (aesthetics), yet still on the "wrong" side (the industrial process).

The search for beauty and simplicity continued with some early attempts to connect these goals with a semi-industrial process. At the beginning of the 20th century, many designers worldwide—among them, Louis Comfort Tiffany and Gustav Stickley in the United States—praised the qualities of taste and simplicity as opposed to the perceived vulgarity and redundancy of most industry. The introduction of electricity provided not only new domestic comforts but also a new visual model of inspiration, the appliance, which was celebrated in the first half of the century by such independent design professionals as Raymond Loewy, John Vassos (fig. 170), Walter Dorwin Teague, Kem Weber, and Norman Bel Geddes.

At that time, as advances in techniques improved its capacity to express lightness and speed, aluminum revealed a new face and introduced new possibilities for design. Its technology and distinct industrial personality had developed through its first applications in the transportation industry, and aluminum was quickly appropriated by designers as one of the materials of choice throughout the spectrum, for domestic and institutional furnishings, means of transportation, and construction materials alike. In particular, the American style called "streamline," for which aluminum was largely responsible, came from observing abstract forms engineered for industrial production, and it dramatically symbolized America's present and future. Oddly inflected by an organic and biomorphic sensitivity, streamline became in the 1930s and '40s the American national style for everyday objects, such as Russel Wright's series of spun aluminum cocktail shakers and trays (ca. 1933) and Raymond Loewy's famous but never-produced pencil sharpener (1934), and then cars, trailers, Loewy's Greyhound buses (1940), typewriters, and vacuum cleaners, like Lurelle Guild's Electrolux model XXX (Highlight 19). Furniture generally escaped streamline's influence, with the exception of the celebrated Airline armchair (1934) by Kem Weber, the curvilinear aluminum table by Frederick Kiesler (fig. 171), and, of course, Warren McArthur's collections (Highlight 13). Many designers, Wright for instance, turned to aluminum after having tried other materials, and adopted it for its ductility and ease of manufacture.

Various political and financial turmoils worldwide, from the Depression in the United States to Fascism in Italy, propelled the search for new materials and moved designers toward a more economical perspective. All over the world, curiously, this step backward helped countries discover the value of their own national design traditions as well as their manufacturing horizons.

Figure 170.
**John Vassos,** designer
(American, b. Romania, 1898–1985)
Radio Company of America,
now owned by Thomson
Multimedia, manufacturer
(American, 1919–present)
RCA Victor Special, model N,
ca. 1937
aluminum, chrome-plated
steel, velvet, and plastic
cat. 1.5

Figure 171.
**Frederick Kiesler**
(American, 1890–1965)
Nesting coffee table, 1935–38
cast aluminum
cat. 3.39

Aluminum, in the countries where it was widely available, suggested new expressive possibilities, ranging from Art Deco and streamline to functionalism. The bauxite abundantly available in Italian territory made aluminum—named *anticorodal* under the policy of autarky—a national material during the closed-border years of Fascism (Highlight 22). A milestone in the innovative exploration of the material was the Bialetti coffeemaker (fig. 172), which marked not only the introduction of a new technology for aluminum fusion but also a new typology of coffeemaker that has remained essentially unchanged to this day. Alfonso Bialetti applied a technique called "shell-fusion," developed in France, and understood that the particular chemical process that aluminum underwent as the coffee boiled would enhance the flavor of the drink.

At this time, one great chair joined the group of objects that constituted a virtual manifesto for the new aluminum era. Walter Gropius and Le Corbusier were among the jurors for a 1938 competition sponsored by the Swiss Parks Authority and won by Hans Coray, a student of Romance languages and self-taught designer. The competition called for new outdoor chairs to furnish Swiss parks. The chairs were to be made of aluminum, the technology for its use being particularly advanced in Switzerland. Coray's chair (fig. 175) became the official seating of the 1939 "Landi," the Swiss National Exhibition (Highlight 21). The fruit of a collaboration between Coray, exhibition architect Hans Fischli, and the Blattmann Metallwarenfabrik manufacturing company, the chair weighed only 6.6 pounds. Its manufacture combined technologies borrowed from the aeronautical industry and the Swiss Federal Railway,

Figure 172.
**Alfonso Bialetti,** designer
(Italian, 1888–1970)
Bialetti Industrie, manufacturer
(Italian, early 1950s–present)
Moka Express coffeemakers, 1999;
designed 1930, manufactured
1933–present
cast aluminum
cat. 3.70

and required a 300-ton drawing press. Its seat was made of aluminum perforated in such a way as to create a chair both very light and exceptionally sturdy.[2] A true industrial product, in typical Swiss spirit, its sensibility was easily translated into many derivative seating designs later produced worldwide. It provides one of the best-realized examples of a cutting-edge and efficient use of aluminum.

After World War II, the need for middle-class housing sparked research into economical ways to build and manufacture both residences and the furnishings and objects within. The "modernity" of design at that time derived from a more rational aesthetic that took into account comfort as well as price and maintenance, and aluminum fit the bill. Manufacturers and designers worldwide began to exploit the technological advances achieved by industry during the war, experimenting with fiberglass and new steel and aluminum manufacturing techniques, and envisioning a new civilian life for many of these materials.

But aluminum, still perceived as the stereotypical material for hospital, military, and school furniture, was not yet accepted by the public as a substitute for wood and other traditional materials in the domestic environment. Its most frequent use at that time was still on a smaller scale. Aluminum was the material of choice for many accessories and appliances, from kitchen equipment to desktop items. In this role, aluminum proved essential to the development of a quintessentially American branch of design—branding. Thanks to its sculptural malleability and the receptiveness of its surface to labels and logos, aluminum became the support and the vehicle for vast branding campaigns, from Coca-Cola—first on its dispensers, then on the cans years later—to Airstream (Highlight 16).

Design is, after all, a search for the perfect balance between means (the available materials and techniques) and goals (for example, a super-light chair, a low-cost steel floor lamp, an affordable sports car). When hands-on techniques were widely replaced by industrial technologies in the first half of the 20th century, the ethical standards of good design established by the Arts and Crafts movement of the 19th century could and often did remain intact. In many successful cases, manufacturers were active players in the design game, in that they owned the means and they supported and addressed the goals. The divide between design and manufacture has since been crossed by many outstanding examples worldwide of companies able to participate in and improve upon the design process by taking an ethical position regarding industrial craftsmanship—for instance, Olivetti, Braun, and Herman Miller. In some cases, like Thonet and Knoll, the designers themselves became entrepreneurs, and the convergence of design and manufacturing processes became even more evident. This is the postwar milieu that provided enough formal and functional innovations to fill an entire encyclopedia on the design process as well as a new chapter on the biography of aluminum.

As we move closer to contemporary times, aluminum was tested in various applications all over the world. Particularly interesting was its new use in the electronics industry, where its intrinsic elegance was celebrated,

Figure 173.
Sony Electronics Inc.
(Japanese, 1946–present)
Walkman, 1979
aluminum and other materials
cat. 1.7

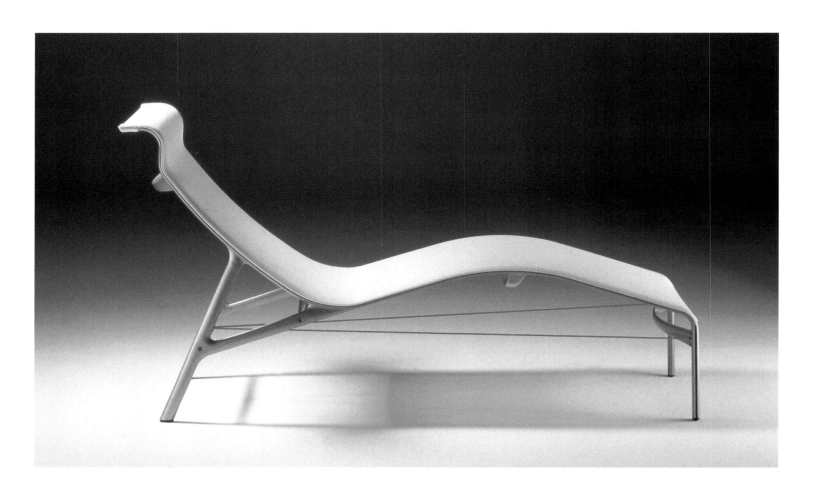

Figure 174.
**Alberto Meda,** designer
(Italian, b. 1945)
Alias Srl, manufacturer
(Italian, 1979–present)
Long Frame, designed 1994,
manufactured 1994–present
die-cast aluminum and
elastic fiber

Figure 175.
Blattmann Metallwarenfabrik AG
advertising photograph of
Hans Coray's Landi chairs, 1959

for instance, by Jacob Jensen in his extensive production of amplifiers, turntables, radios, and all sorts of audiovisual equipment for Bang & Olufsen, as well as by the in-house design team at Sony, which used aluminum to screen the loudspeakers and to support the controls of the first transistor radios in the 1950s. The Walkman (fig. 173), introduced in 1979, also made of aluminum, was another milestone. In a lecture at the 1998 International Design Conference in Aspen, Colorado, writer Malcolm Gladwell declared the Walkman responsible for a general improvement in the fitness of Americans, as it made running on treadmills and lifting weights bearable, even fun. The Walkman, nonetheless, represents far more. It obliterated decades of studies and measurements of the human body and the space it inhabits. Its heads and buttons introduced a customized, compact, personal environment, a portable bubble of individuality. The Walkman is testimony to the profound but silent revolution in the design of private spaces, and aluminum enabled such a revolution.

Unlike plastics, aluminum has never saturated the market. A comparison with plastics can help us understand aluminum's subtle role in material culture. Materials and techniques are not only the most useful tools and sources of inspiration for designers, they can also be the most efficient means of narration for a design writer and curator. From a curator's viewpoint, plastics can easily tell the greater part of the history of design in this century. Shifts in perspective could provide many different narrations of the same history. An encompassing historical exhibition would trace the design path of plastics from their imitation of natural materials such as tortoise-shell and ivory in the 19th and early 20th centuries, to the freedom of expression provided by thermosetting polymers. A targeted sociological design survey might instead follow plastics' role in the home, from tabletop objects in the 1930s to functional tools in the 1950s, on to their use in the suffocating global environments of the 1960s and rejection in the 1980s, and finally their recent reintroduction as elegant and durable materials. An explanation of the act of design could disclose the decisive role of materials in the creative and manufacturing process at any chosen moment of history. Plastics, and their derivative composites, were so pervasive that they often replaced other materials, certainly aluminum, in many contexts.

Aluminum, by never overwhelming the design landscape, has remained a material of accents, used to enhance rather than lead the aesthetic revolution that drove the first two centuries of industrial design. Its seamless history is that of a great character actor. In our contemporary visual culture, so attentive to detail and subtext, aluminum can at last shine in all its complexity and richness. While the aeronautical and sports industries are exploring new alloys and processes to overcome natural performance weaknesses, in design at large aluminum has also recovered its initial formal expressivity. It is sometimes used in the guise of cast iron to produce sculptural pedestals for chairs and tables when the main structure is made of other materials—plastics, fiberglass, or even fiberboard. Such a use is found in the work of Stefan Lindfors, Aaron Lown, and Ali Tayar, among others (figs. 176–177). Often, the mark of sand from the cast is left exposed to show a nostalgia for preindustrial times and a continuity with the past, albeit in an exquisitely contemporary form. In Marc Newson's Lockheed Lounge (Highlight 37), we sense

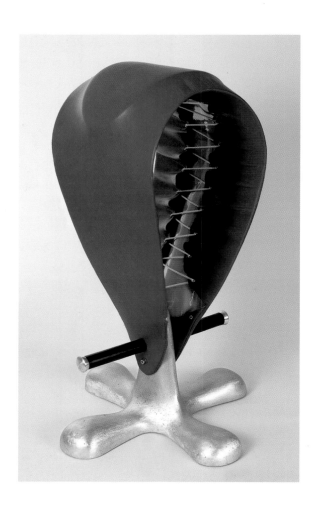

Figure 176.
**Ali Tayar,** designer
(American, b. Turkey, 1959)
Parallel Design Partnership,
manufacturer (American,
1992–present)
NEA Table 1, 1994
recycled molded particleboard,
cast aluminum, and glass

Figure 177.
**Aaron Lown** (American, b. 1968)
Hi Ho Stool, prototype, 1994
vacuum-laminated fiberglass,
urethane foam, machined
aluminum, sand-cast aluminum,
rubber, and leather
cat. 5.36

a yearning for the early days of aircraft technology. In this iconic and sculptural piece, thin-walled aluminum sheets are attached with blind rivetings to a fiberglass body, to mimic the glimmering skin of an early airplane fuselage.

Not all contemporary furniture design in aluminum has a retro feeling, of course. The metal has also been used in structural and experimental fashion, for instance in the work of Alberto Meda, an engineer and designer who has produced some of the most interesting and innovative domestic structures of the past fifteen years. "The technology and the new materials are a large warehouse of creative suggestions which, when looked at with interpretive ability, go beyond their strictly technical performance," Meda writes in an article explaining his Softlight chair (fig. 178), made of advanced composites.[3] While conducting a parallel search in the universe of carbon fiber and other composites, Meda has been able to isolate the inner, biomorphic soul of aluminum. If steel is muscle, in Meda's work aluminum is tendon and bone, fragile and powerful at the same time, and certainly structure-defining.

In Meda's 1994 Frame series of seating furniture (fig. 174), for example, the apparently seamless continuity between the elastic fiber seating and the aluminum structure is achieved by taking advantage of the latest extrusion and casting technologies and design ideas appropriated from the sailing world, such as the cables that hold fibers in tension. These seats are technological masterpieces in the research and attention to detail they demand. The outcome, nonetheless, does not rely on technology to define a formal style but, rather, on its use to achieve fluidity and unity of shape and structure. "Paradox: the more complex the technology, the more it is suitable for the production of objects for simple use, with a unitary image, almost organic," Meda asserts.[4]

Titania, a formally striking aluminum lamp designed by Meda with Paolo Rizzatto in 1991, is an expressionistic celebration of the power of technology (fig. 179). Light blades and ellipses are shear-cut from aluminum sheets composed in a complex structure that recalls airplane wings. Red, purple, yellow, blue, and green silkscreened polycarbonate filters can be inserted in slots amidst the blades to obtain colored lateral light, while the light directed at the table and ceiling remains white. The lamp, suspended by seemingly immaterial wires from the ceiling, floats in midair like a quirky spaceship.

Today, technology offers designers a new and exhilarating freedom. Contemporary designers have learned to play with it as in a jazz improvisation, with an elegant nonchalance that can be acquired only through years of strenuous technical exercise. Agile and knowledgeable, designers have learned to give concreteness to their best ideas. Inspired by a humanity undergoing rapid change and evolution, they are finding

Figure 178.
**Alberto Meda,** designer
(Italian, b. 1945)
Alias Srl, manufacturer
(Italian, 1979–present)
Softlight chair, 1989
honeycomb aluminum, molded
carbon fibers in epoxy-resin matrix,
and elastic fiber
cat. 5.39

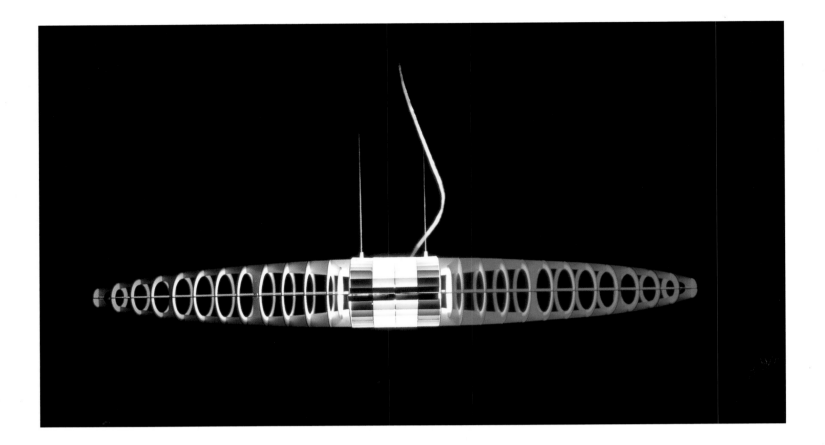

new ways to address the myriad issues that design can master, and appropriating new methods of intervention as they go. In the current design climate, based on stylistic freedom and individual interpretations of the tasks at hand, all of aluminum's qualities are called upon simultaneously. Aluminum is often the material of choice not only because of its technical possibilities but also because of its nature, expressive of youth, purity, and cleanliness, and its capacity to evoke the new and nostalgic at the same time. Aluminum is again among the protagonists of much formal and structural research.

A stereotype can provide powerful insights into not only the characteristics of the subject it describes, but also the expectations inherent in its cultural context. Throughout its history, aluminum has embodied many of the expectations on our wish list for the material world, making it one of the most emblematic materials of our time.

First and foremost, aluminum embodies lightness. "By the year 2040," argues Dr. G. J. Wijers, Dutch minister of economic affairs, in a volume dedicated to the subject of lightness, "the world's population will have approximately doubled. People living then will be enjoying an average level of prosperity that is far higher than today. Without changes in production and consumption, that would lead to an unacceptable pressure on the environment.... In a sustainable economy, the guiding principle is 'the lighter the better.'" [5] Lightness is the leitmotif of much celebrated contemporary design, the obsession of much research in materials technology, and a feature of almost instant appeal to consumers. Aluminum's identification with lightness is today very different than in the 1930s. Rather than merely suggesting possible developments in global weight reduction, aluminum technology has fully achieved several of them. Bicycle design—such as that of the Italian Cinelli brand, a favorite of bicycle aficionados worldwide—was among the first fields of experimentation outside a military context. Since the 1960s, aluminum and its

Figure 179.
**Alberto Meda,** designer
(Italian, b. 1945)
**Paolo Rizzatto,** designer
(Italian, b. 1941)
Luceplan SpA, manufacturer
(Italian, 1978–present)
Titania lamp, designed 1991,
manufactured 1991–present
aluminum and polycarbonate

alloys have been tested in Cinelli's headquarters—its Milan production facilities are more akin to surgical rooms than a factory—in attempts to reduce, even infinitesimally, the weight of bicycle bodies and achieve enhanced performance. The results are design masterworks, their beauty relying on a faithfulness to the material's nature and possibilities (fig. 180). Michael Burrows's Windcheetah T.I. of 1992 (fig. 181), an extreme application of aluminum in human-powered vehicles, carries aluminum's structural possibilities even further by employing one aluminum alloy for its hand-cast lugs and another, sturdier alloy for its tubular frame.

British designer Ross Lovegrove is currently exploring civilian applications of an extraordinarily light and spongelike aluminum filter for hydrogen tanks, which he imported from Japan. Another good example of the striving toward lightness is the much celebrated Audi A8 car, introduced in Europe in 1994. Its all-aluminum body, made of seven different aircraft-grade alloys, is 40 percent lighter than a comparable steel body (Highlight 42). The material is nonetheless still less accepted than steel because, unlike many bicycle shops, the average auto body shop is not equipped to repair damaged aluminum. As always, introducing a new material, or a new use for a traditional material, in the mass market requires considerable investment and strategizing.

A second stereotypical characteristic is that aluminum well represents design's ambivalent and nonideological relationship with technology. As previously described, aluminum can be both high tech and low—sometimes within the same designed item—pliable and compliant in the designer's hands. Spun, cast, extruded, stamped, anodized, used in alloys and in ceramic composites, aluminum is "perhaps one of the most versatile materials," according to designer Jasper Morrison, "in the sense that so many processes can be applied to it. And still it retains an aura of space-age modernity despite developments in plastics and composite materials."[6]

Introduced as a fine arts material and further developed by such heavy hitters as the military and aerospace industries, aluminum lends itself to both the subtle textile fantasies of Junichi Arai (Highlight 36) and Ron Arad's assertive Tom Vac stacking chair (Highlight 35), a monument to stamped-out mass production. Shiro Kuramata, the late designer and the greatest poet of materials, made aluminum sublime in accents and details that exploited its industrial and flexible qualities. He used aluminum, for example, in the pink legs of the Miss Blanche chair (fig. 184) to ground the cloud of red roses embedded in translucent acrylic, and in the small knobs of many dressers. Similarly, Italian designers Afra and Tobia Scarpa have used aluminum, anodized and varnished to achieve the effect of a luxurious car body, in a series of multicolored dishes that are simultaneously exquisitely industrial and as precious as jewelry (fig. 183).

As a consequence of its adaptability to industrialized manufacturing processes, aluminum can indeed signify mass production. Certainly in the Tom Vac chair, but also in other iconic designs such as Michele De Lucchi and Giancarlo Fassina's Tolomeo lamp (fig. 186), or Rizzatto's Costanza and Costanzina lamps (fig. 187), aluminum is the material of repetition and economy. The Tolomeo, the integral aluminum lamp that represents the apotheosis of the use of this metal in the late 1980s—and

Figure 180.
Cinelli, now a division of Gruppo SpA
(Italian, 1948–present)
Starship, 2000
aluminum alloy 6000 and
other materials

Figure 181.
**Michael Burrows,** designer
(British, b. 1943)
Seat of the Pants Co., Ltd.,
later renamed Advanced Vehicle
Design, Ltd., manufacturer
(British, 1993–present)
Windcheetah T.I. HPV (Human
Powered Vehicle), designed 1992,
manufactured 1995
aluminum, titanium, and carbon

Figure 182.
**Ali Tayar,** designer
(American, b. Turkey, 1959)
Parallel Design Partnership,
manufacturer (American,
1922–present)
Wernerco, extruder of aluminum
elements (American, 1922–present)
Ellen's Brackets, 2000; designed 1993,
manufactured 1993–present
extruded aluminum
cat. 5.33

that to this day is a best-seller for its manufacturer, Artemide—is a catalogue of diverse industrial aluminum technologies, containing aluminum parts that are either stamped, formed, or cast.

Similarly, Ali Tayar's aluminum brackets, initially produced by the New York–based designer in a small series, are as economically suited as window-frame profiles for production on a very large scale (fig. 182). The formal idea came from a force-flow cantilever diagram, and Tayar applied aluminum so that its parabolic curve, touching zero at the tip of the cantilever, would reach its maximum at the point of support. The extrusion process, which guarantees complete freedom along two axes, is one of the simplest and most common ways in which aluminum has entered the manufactured world, and was the technique of choice in this case. Aluminum's mass appeal has recently been celebrated in the new Hudson chair by Philippe Starck for Emeco (Highlight 34), the Pennsylvania-based manufacturing company that specializes in mass-produced aluminum furniture.

In the list of aluminum's design stereotypes, lightness, versatility, and mass appeal are followed by "hygienic" and "cool." Aluminum, the ubiquitous material for hospital beds and school furniture, is today often employed to control the visual temperature in many a brand line, especially in stationery and cosmetics packaging. The Halliburton attaché case, initially introduced in 1938 and brought to its current design in 1946 (fig. 191), was one of the first examples. About ten years ago, the craze for aluminum carrying cases exploded. The briefcase, rugged yet delicate, gave birth to a whole family of luggage (fig. 190), and the inevitable dents and scratches from early 1990s airport carousels were regarded almost as warrior's scars.

The Halliburton cases, certainly heavier than their equivalents in leather or plastic, were chosen by many simply for their look. Muji, the Japanese chain store famous for its minimal, chic, and restrained products, and Stila, the oversimplified, oversophisticated cosmetics line, provide excellent examples of aluminum's suggestive power as well as changed expectations in matters of style. Muji's desk items (fig. 189), for example, and Stila's lip-gloss packaging, which resembles old-fashioned gouache tubes, exploit aluminum's blue hue to enhance the essential coolness of the look, as well as its nostalgic call for a cleaner, leaner future.

Softness and formal malleability are other features of aluminum that have drawn designers to connect its use to iconic shapes. As Russel Wright and Raymond Loewy well knew, aluminum is easier to work than any other metal and therefore can be much more formally expressive. From Gijs Bakker's bracelets at the end of the 1960s (Highlight 32), to the work of several artists like Donald Judd (fig. 188) and Andrea Zittel, recent work in aluminum can be exquisitely sculptural. Karim Rashid chose aluminum for his 1999 *Digitalia Time Capsule,* the shiny and formally suggestive metal body of which encases an LED screen featuring a continuous display of neologisms from the 1960s to the end of the millennium. Fernando and Humberto Campana, the São Paulo brothers who have recently gained recognition for their peculiar combination of traditional techniques and advanced materials, have treated aluminum in the guise of bamboo in their *Sculpture Screen* (fig. 169), while Belgian hyperminimalist Maarten Van Severen has taken the metal to its extreme in a simple

Figure 183.
**Afra Scarpa,** designer
(Italian, b. 1937)
**Tobia Scarpa,** designer
(Italian, b. 1935)
San Lorenzo Srl, manufacturer
(Italian, 1970–present)
Dishes, 2000; designed 1992,
manufactured 1992–present
anodized and varnished aluminum
cat. 5.47

Figure 184.
**Shiro Kuramata,** designer
(Japanese, 1934–1991)
Ishmaru Co., Ltd., manufacturer
(Japanese, 1969–present)
Miss Blanche, designed 1988,
manufactured 1989–98
acrylic resin, paper flowers,
and aluminum
cat. 5.38

chair, shaped like an apostrophe, in which aluminum is meticulously crafted to provide maximum bounce and curvature (fig. 185).

Last in this list of contemporary stereotypes is "controversial." "I was thinking," says Droog Design veteran Hella Jongerius, "nobody is doing anything with aluminum. Why? Well, it's not the period to do something in metal: it's too cold, hard, anonymous. And it has this smell of the eighties, with all the shining aluminum design stuff for yuppies' desks." Jongerius, whose Midas touch with materials has become a formula for success, nonetheless also fell victim to aluminum's coolness: "On the other hand, I just bought at Muji in London a very smooth and matte surface. This soft look of aluminum is new and human."[7] A material's ability to heat up a discussion is a sure sign of its validity in today's design world, which is opinionated and determined to give meaning to all choices, even the most preliminary ones, of goals and means.

The birth of a mutant material begins with design, with the first act of transformation of matter into a usable material: earth powders into glass and ceramics, oil into thermoplastic pellets, wheat into flour. The true mutant character of materials lies in a striving toward perfection. The obsession focuses on intrinsic weaknesses that must be overcome. Ceramics must cease to be fragile to become the perfect heavy-duty material. Plastics must become more stable, durable, and aesthetically appealing so as to become the natural answer to the depletion of natural resources. Fibers must be added to resins to produce isotropic, light, resistant bodies. Aluminum is a source of inspiration for much of this research inclined toward perfection.

The best contemporary objects are those whose presence expresses history and contemporaneity; those that exude humors of the material culture that generated them, while at the same time speaking a global language; those that carry a memory and an intelligence of the future; those that are like great movies in that they either spark a sense of belonging—in the world, in these times of cultural and technical possibilities—while they also manage to transport us to places we have never visited. The best contemporary objects are those that express consciousness by revealing the reasoning and processes that led to their making.

Aluminum helps us understand the importance of material culture in the future evolution of the constructed world. Served by technological progress, local culture has in recent decades proved to be, for design and architecture alike, the safest and most efficient way to move beyond modernism without giving up the great qualities of modern design. Shiro Kuramata was among the first to exemplify this notion. Kuramata took the most established rules of modernist design and sifted them through his Japanese sensibility. He took a classic black-and-white cubic dresser and deformed it gently, by having it wave on its own axis. He took white bookshelves formed in a rigid grid and varied the rhythm of the grid within the piece. By attacking only one of the variables in the modernist equation, rather than many all at once, he created surprise and enlightenment, and he did so by learning from his local tradition.

In recent times, the evolution of the role of technology, high and low joined
in the pursuit of new concepts of beauty and functionality, has brought many
local cultures to the forefront in unexpected ways. Many countries whose material
traditions are based on craftsmanship and whose economies are based on necessity,
like Brazil, are now being looked at as new paradigms in architecture and design.
Available materials are used in harmony with their capabilities, according to what
Arthur Pulos calls "the principle of beauty as the natural by-product of functional
refinement," a timeless principle that has often been used to define quality in design.[8]

A clear view of the future of design can be achieved by considering its present, which is filled with expectations and undergoing continuous evolution. The
design for the new century is being made today. "More possibilities" is today's motto.
It works both for the impalpable Internet and for the very palpable materials of
design. More possibilities pose a challenge, in that choice becomes a greater responsibility. The mutant, or should I say "virtual," character of contemporary materials
calls for an increasingly rigorous and conscious design approach. Designers are
already responding, spontaneously, to this call and producing objects that are more
honest and relevant.

The present moment is peculiarly similar to the period after World War II.
These are spiritual times, marked by a renewed attention to domestic life and fueled
by concerns about the environment and a strong political consciousness worldwide.
Design trends are often accurate reflections of social change, and the economy
and sensibility that envelop the world today after an age of excess are very strong

Figure 185.
**Maarten Van Severen**, designer
(Belgian, b. 1949)
Maarten Van Severen Meubelen,
manufacturer (Belgian,
1988–present)
Low Chair, 1999; designed 1993–95,
manufactured 1995–present
aluminum
cat. 5.24

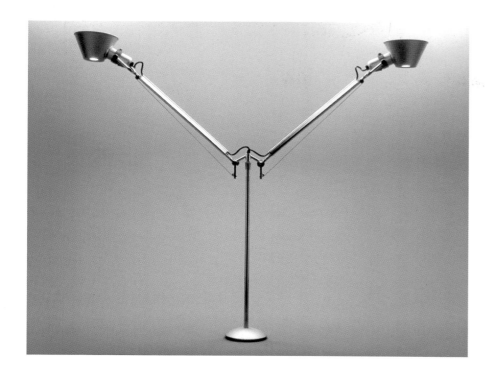

Figure 186.
**Michele De Lucchi,** designer
(Italian, b. 1951)
**Giancarlo Fassina,** designer
(Italian)
Artemide, Inc., manufacturer
(Italian, 1959–present)
Tolomeo lamp, designed 1987,
manufactured 1987–present
polished and anodized aluminum

Figure 187.
**Paolo Rizzatto,** designer
(Italian, b. 1941)
Luceplan SpA, manufacturer
(Italian, 1978–present)
Costanzina lamp, designed 1986,
manufactured 1986–present
aluminum and polycarbonate

Figure 188.
**Donald Judd,** designer
(American, 1928–1984)
Janssen, manufacturer
(Dutch, 1957–present)
Armchair, 2000; designed 1984,
manufactured 1984–present
anodized aluminum
cat. 5.43

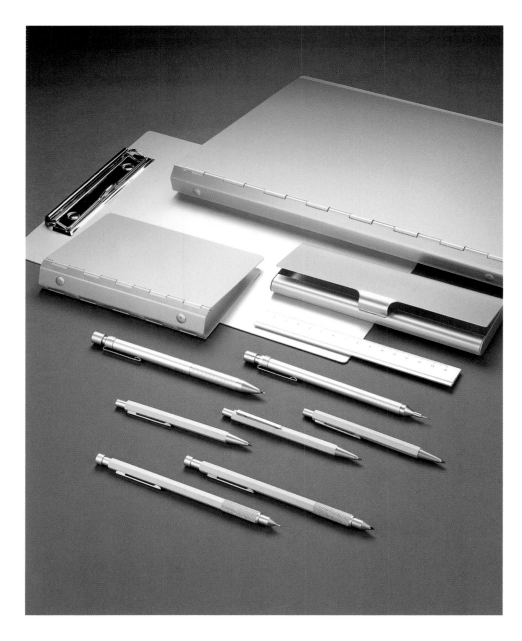

Figure 189.
Ryohin Keikaku Co., Ltd.,
manufacturer
(Japanese, 1979–present)
Muji, retailer
(Japanese, 1983–present)
Group of objects, 2000
aluminum
various sizes
cat. 5.49

(opposite)
Figure 190.
Zero Halliburton
(American, 1938–present)
Luggage and attaché cases, 2000
aluminum, leather, polyester,
and neoprene

Figure 191.
Advertisement for Halliburton,
later renamed Zero Halliburton
*Fortune* (December 1946): 29

forces indeed. Morality, sometimes even moralism, is a recognizable feature of many contemporary objects. In good recent design, ethics are as important as aesthetics. All in all, contemporary design is frequently experimental in its use of materials and often inspired by genuine necessity. Even so, it sustains elements of surprise and deep intellectual beauty because it relies more on invention and reduction than on the elaboration of previous styles. It is a good time for design.

Today, design is self-assured and assertive, and designers are experimenting with all the expressive resources of materials. In this context, aluminum is being given a wider and wider range of applications in which to articulate its complex personality. In some cases, it has become the material of nostalgia and memory: in Masuda's boxes and Rashid's *Time Capsule*, for example. Yet aluminum cannot help but embody the optimism that accompanies a reorientation toward a better, cleaner, simpler world. It is the material that best expresses nostalgia for the future.

## Notes

1. Paola Antonelli, *Mutant Materials in Contemporary Design* (New York: Museum of Modern Art, 1995).

2. Alexander von Vegesack, Peter Dunas, and Mathias Schwartz-Clauss, eds., *100 Masterpieces from the Vitra Design Museum Collection* (Weil am Rhein, Germany: Vitra Design Museum, 1996), 34.

3. Alberto Meda, "How Design and Technology Interact," *Domus*, no. 761 (June 1994): 76–79.

4. Ibid.

5. Dr. G. J. Wijers, preface to Adriaan Beukers and Ed van Hinte, *Lightness: The Inevitable Renaissance of Minimum Energy Structures*, 2d ed. (Rotterdam: 010 publishers, 1999).

6. Interview with the author.

7. Interview with the author.

8. Arthur J. Pulos, *American Design Ethic: A History of Industrial Design to 1940* (Cambridge, Mass.: MIT Press, 1983), 7.

# Aluminum
## by design

## Highlights

Charles Eames
(American, 1907–1978)
Ray Eames
(American, 1912–1988)
*Solar Do-Nothing Machine*, 1957
anodized aluminum

Jean-Auguste Barre (French, 1811–1896)
Eagle, designed 1860
gilded aluminum
cat. 2.7

Charles Rambert, designer
(French, active Paris after ca. 1848)
Honoré-Séverin Bourdoncle, maker
(French, 1823–1893)
Baby rattle for the Prince Imperial, 1856
aluminum, gold, coral, emeralds,
and diamonds
cat. 2.4

F. Desbœufs, engraver
(French, active mid-19th century)
Medal in original box, 1856
aluminum, wood, morocco leather,
velvet, silk, and gold leaf
cat. 2.6

(opposite)
Charles Christofle, designer
(French, 1805–1863)
Charles Christofle et Cie, manufacturer
(French, 1830–present)
Centerpiece, 1858
aluminum, silvered-copper alloy,
and gilded bronze
cat. 2.5

Intellectual curiosity and the promise of national glory inspired Napoleon III's support of technical entrepreneurship. And Henri Sainte-Claire Deville's development of a chemical process for producing aluminum in usable quantities certainly qualified as a vehicle for the advancement of the Second Empire. A number of aluminum objects from the 1850s are directly connected with Napoleon III as either a donor or recipient. Although aluminum artifacts were exhibited at the 1855 Paris Exposition, one of the earliest known is the baby's rattle commissioned by the Emperor to commemorate the birth of his son, the Prince Imperial, in 1856. Two years later, silversmith Charles Christofle, another beneficiary of royal largesse, presented Napoleon III with a table centerpiece composed of aluminum figures of five children, symbolizing prosperity, mounted on a gilded bronze and silvered-copper alloy base. The inscription celebrates "the help and encouragement" the Emperor "has brought to the dedicated work of the learned Henri Sainte-Claire Deville in the making of aluminum."

Although the first aluminum objects were small—for instance, medallions and luxury or novelty items—Napoleon III immediately recognized the advantages of this light metal in the military arena. Use of aluminum would dramatically reduce the weight of equipment such as spurs, buttons, sword handles, saber sheaths, helmets, and the imperial eagles mounted atop flagpoles carried into battle. In 1860, a new model for the imperial eagle was approved. Made from gilded aluminum, it looked no different from the bronze version, but its weight was reduced from 2.4 to 0.9 kilograms. From 1861 onward, the aluminum eagle replaced the bronze version among the French regiments. In spite of this initial advance, aluminum had little impact on the military in the 19th century, in France or elsewhere. But 20th-century warfare would eventually confirm Napoleon's faith in the metal in ways he could not have foreseen.

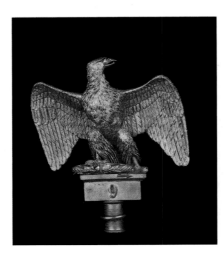

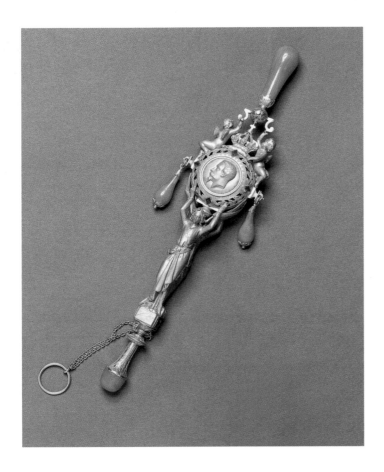

Probably French
Suite of jewelry (brooch, earrings,
and cuff studs), ca. 1860
aluminum and gold
cat. 2.20

French
Bracelet, ca. 1858
aluminum and gold
cat. 1.1

(opposite)
Attributed to **Honoré-Séverin
Bourdoncle** (French, 1823–1893)
Bracelet, ca. 1858
aluminum and gold
cat. 2.21

Elaborate aluminum bracelets and brooches were produced
as luxury items during the late 1850s and early 1860s by many
notable Parisian jewelers. At this time, aluminum was always
combined with other, often more expensive, materials such
as gold or gemstones, usually in a way that required tremendous time and skilled workmanship. For example, aluminum
and gold could not be easily soldered together but required
riveting, and the rivet heads thus had to be made into a
design feature. When it wasn't treated three-dimensionally,
aluminum was often engraved to create texture and the
play of light on otherwise solid, flat surfaces.

One of the most famous jewelers to produce aluminum
objects was Honoré-Séverin Bourdoncle, a successful chaser
who was awarded silver medals at the French Expositions
of 1855 and 1867. Among his clientele was Napoleon III
(Highlight 1). His bracelets, inspired by 16th- and 17th-century
designs, are sculptures in miniature.

The heyday of luxury aluminum jewelry, from about 1856
to 1865, came to an end when refinements in French production processes caused a decline in the price of aluminum.
At this point, aluminum lost its cachet as a technically
innovative, precious metal and began to be used in the
manufacture of more everyday objects.

**Sir Alfred Gilbert,** designer
(British, 1854–1934)
Compagnie des Bronzes, manufacturer
(Belgian, active 1853–1935)
*St. George,* 1899
cast aluminum
cat. 2.62

Photograph of the Shaftesbury
Memorial, Piccadilly Circus,
London, ca. 1900

(opposite)
**Sir Alfred Gilbert** (British, 1854–1934)
Cast by George Mancini (b. 1903)
at Morris Singer Foundry
(British, 1848–present)
*Eros,* 1985 cast of original (1893) on the
Shaftesbury Memorial, Piccadilly Circus
aluminum
cat. 2.63

Corrosion-resistant and easier to cast than bronze, aluminum is an ideal medium for outdoor sculpture. By the 1890s, with the increase in the supply of aluminum, it became an obvious application for the metal. The first aluminum public sculpture was made by Sir Alfred Gilbert. In May 1886, Gilbert was commissioned to design a memorial to the British politician and reformer Lord Shaftesbury. Located in Piccadilly Circus, London, the memorial consists of a large base with an iconography of marine forms topped by a fountain cistern. An addition to Gilbert's original conception is the aluminum statue of Eros, a representation of selfless love later introduced into the sculptural scheme. Gilbert collaborated with metallurgist William Chandler Austen Roberts on the casting of the memorial's various elements and also kept in close contact with the Society of Arts, which sponsored lectures on developments in aluminum and metallurgy. He chose aluminum not for its lightness but for its color, to contrast with the reddish-gold and green of the bronze base and cistern. As Richard Dorment has written, "he was aiming at a precious jewel-like ensemble, rich and intense, but also bold, to stand out in the English fog."

According to *Aluminium and the Non-Ferrous Review,* when Eros was dismantled for cleaning in 1931, conservators were astounded to discover that no deep corrosion had occurred, in spite of forty years of exposure to urban pollution. A layer of corrosion about one-sixteenth of an inch thick was removed, leaving a perfectly smooth surface. In 1984–85, Eros was again taken down for restoration. At that time, the original plaster molds were restored as well, and a limited edition of this famous statue was cast from these restored molds.

Gilbert again used aluminum for the tomb of Prince Albert Victor, duke of Clarence, in St. George's Chapel at Windsor Castle. Among the statues surrounding the tomb, one of St. George, made in 1898, is clad in cast aluminum armor. The following year, Gilbert produced two additional statues of St. George cast entirely in aluminum. The most prominent British sculptor of the late 19th century, Gilbert was also a goldsmith, which may explain his interest in modern metals and new metalworking techniques.

**Talwin Morris** (Scottish, 1865–1911)
Buckle, ca. 1900
aluminum set with foil-backed pastes
cat. 2.37

**Margaret Macdonald** (Scottish, 1865–1933)
**Frances Macdonald** (Scottish, 1874–1921)
Picture frame, 1897
aluminum and oak
cat. 2.36

(opposite)
**C. F. A. Voysey** (British, 1857–1941)
Clock, ca. 1900, designed 1895
aluminum and copper
Courtesy The Birkenhead Collection
cat. 2.38

For its most committed exponents, the Arts and Crafts movement—which encompassed architecture, furnishings, and every aspect of domestic design—was not merely a style but a philosophy of life. Its leading figures advocated handcrafted over machine-made objects, the use of indigenous materials and forms of decoration, simplicity, and truth to materials. One highly individualistic manifestation of Arts and Crafts emerged from the Glasgow School of Art and centered around Charles Rennie Mackintosh and his circle, which included the sisters Frances and Margaret Macdonald. In addition to their commercial and often innovative graphic projects, the Macdonald sisters made watercolors, gesso panels, embroidery, jewelry, and metalwork, including candlesticks, mirror and picture frames, and sconces. The linear quality of their graphics informed the decorative motifs of their objects, as seen on the aluminum frame illustrated here. Although aluminum is not a key material in their work, its use here represents a sensible choice. Not only was it cheaper and easier to work than silver, it could also be considered a Scottish material, since the main factories of the British Aluminium Company Limited (founded in 1894) were located in Scotland. Talwin Morris, another member of the Mackintosh circle, produced work of beaten metal—buckles and brooches in copper, silver, and aluminum. Although his reasons for using aluminum may have been similar to those of the Macdonald sisters, he patinated the surfaces to resemble copper.

A major figure of the Arts and Crafts movement in England was the architect and designer C. F. A. Voysey. A master of decorative detail, Voysey attended to every aspect of his projects, particularly the metalwork, designing door and window handles, ventilator grilles, fireplace surrounds and tools, and even hinges and pulls. His interest in metalwork and relationship with metal fabricators may have spurred him to use aluminum for a number of clocks and inkwells. In 1895, Voysey produced a design for an elaborately painted wooden clock with conventional Roman numerals on the face, which by 1901 he had replaced with the words *tempus fugit* ("time flies"). He produced a small quantity of clocks in both undecorated wood and aluminum based on the highly architectural form of the painted case.

Eastman Kodak Company
(American, 1892–present)
8 x 10 flatbed camera with 10-inch
wide-field Ektar lens, date unknown
aluminum and other materials
cat. 2.47

Aitchison & Co. (British, 1889–present)
Folding binoculars in original case,
ca. 1900
aluminum, glass, leather, and metal
cat. 2.42

Illustration of soldier wearing
Aitchison & Co. folding binoculars
*Aluminium and Electrolysis*
(January 1897): 92

(opposite)
**George Blickensderfer,** designer
(American, d. 1919)
Blickensderfer Manufacturing Co.,
manufacturer (American, active
1893–1915)
Featherweight Blick typewriter, ca. 1894
aluminum, iron, plastic, rubber,
copper, and felt
cat. 2.43

Probably Swiss
Surgical instruments in original case,
ca. 1890s
aluminum, leather, and other materials
cat. 2.44

Aluminum has been used for precision instruments since the earliest days of Deville's production. In fact, the first aluminum object appears to have been the balance arm from a scale inscribed with the date 1855 (fig. 80). Aluminum's qualities enhance the performance of precision instruments. Moving parts made from aluminum react more quickly than parts made from heavier metals. As a nonmagnetic material, aluminum is ideal for compass boxes. It takes on atmospheric temperature faster than any other metal, so a meteorological thermometer with an aluminum back responds more quickly than one with a silver-plated brass back. Its light weight is advantageous for portable or handheld equipment, such as surveyors' sextants and theodolites, typewriters, surgeons' instruments, binoculars, and cameras.

In the 1890s, Kodak produced a folding pocket camera with its aluminum body concealed under leather. The camera was so popular that Kodak marketed several larger models, again with a concealed aluminum body, in 1900. Aluminum was also used for plate holders and other parts. Ansel Adams described his experience photographing Yosemite National Park circa 1950: "I was using my Kodak metal 8 x 10 view camera, of aluminum construction, designed as a replica of their standard wooden flatbed camera. It once belonged to the explorer Louise Boyd, who funded and directed several expeditions to Greenland." Boyd's first expedition was in 1926, and the camera may date from that time.

Aluminum opera glasses, an early application of the metal, were as common as those made of brass and nickel by the turn of the century. Several companies also made binoculars for military use. Aitchison & Co. produced an ingenious type of pocket binocular that folded down to a thickness of 1⅜ inches and weighed about four ounces. The firm also devised a way of freeing the user's hands by attaching the binoculars to a headrest consisting of an aluminum plate with straps. "The construction of the head-rest," reported *Aluminium and Electrolysis* in 1897, "gives an unimpaired view of the note-book from beneath its lower edge; so that sketching or recording, as well as watching any distant events, may be carried on simultaneously; and the user is able to walk about, to ride on horseback, or to cycle."

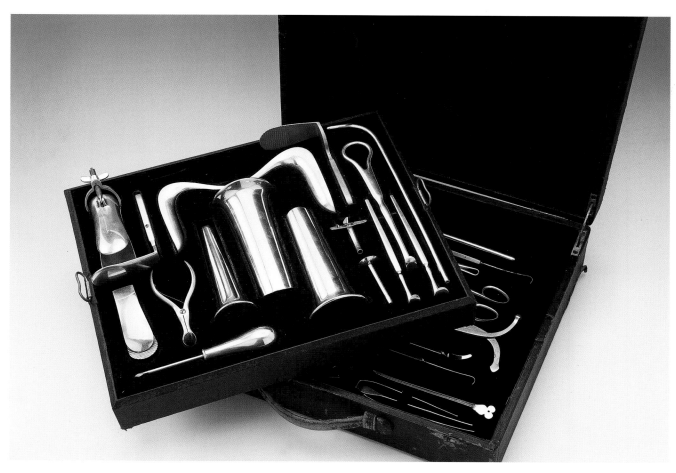

Otto Wagner (Austrian, 1841–1918)
Lamp from the Austrian
Postal Savings Bank, Vienna, 1904–06
aluminum and other materials

Otto Wagner, designer
(Austrian, 1841–1918)
J & J Kohn, manufacturer
(Austrian, 1868–1923)
Die Zeit chair, 1902
beech, aluminum, tape, and fabric
cat. 2.71

(opposite)
Otto Wagner (Austrian, 1841–1918)
Interior of Austrian Postal Savings
Bank, Vienna, 1904–06

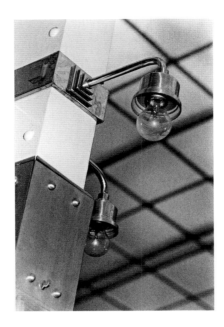

If a single person could be identified as the first proponent of aluminum as a modern material, it would be the Viennese architect and designer Otto Wagner. A pioneer in his field, Wagner advocated straightforward, economical construction and a complete understanding of the purpose of the building or object, down to its smallest details. These principles governed his innovative use of many nontraditional materials, but particularly aluminum.

Wagner incorporated aluminum into several projects, but his first extensive use of the metal occurs in the facade and interior of the news agency Die Zeit (fig. 97), which art historian Kirk Varnedoe has described as "an aggressively untraditional statement wholly appropriate to an office that trafficked in contemporary events." As a working space, the agency's furniture had to withstand wear and tear. Chairs were protected by adding aluminum strips to the arms and shoes to the legs—the two parts of a chair most susceptible to damage. Primarily practical, aluminum also served as decorative articulation and provided a striking color contrast in an otherwise quite minimal linear design.

Wagner's next major building project to feature aluminum, the Postal Savings Bank, has become a landmark of modern architecture (fig. 96). An impressive hierarchy of materials and structure is evident throughout the building, but in the banking hall—the primary public space, which remains today very much as it was in 1906—marble, glass, linoleum, and aluminum are combined in an airy, spacious interior designed for durability and ease of maintenance. For example, the remarkable aluminum warm-air blowers, which seem to belong to a future generation of modernist forms (fig. 1), were designed to keep the glass roof free of snow. With this building, Wagner produced an innovative and harmonious symbiosis of old and new materials and, in the process, situated aluminum well and truly in a forward-looking, 20th-century context.

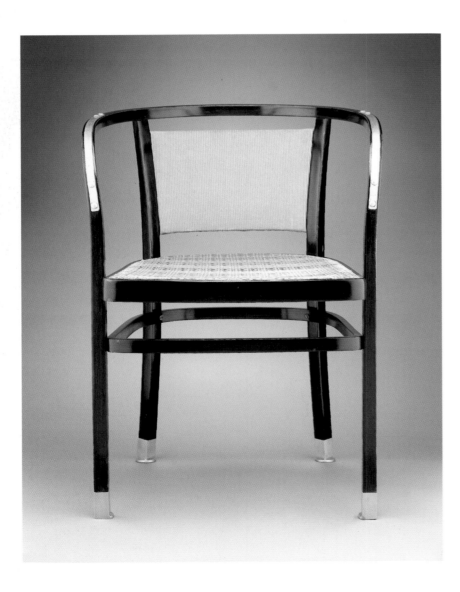

**René Lalique,** designer
(French, 1860–1945)
Roger & Gallet, manufacturer
(French, 1862–present)
Cosmetics boxes, 1922
stamped aluminum and lacquer
cat. 3.55

**René Lalique** (French, 1860–1945)
Prototypes for boxes, 1922 or earlier
stamped aluminum and lacquer
cat. 3.53 and 3.54

(opposite)
**René Lalique** (French, 1860–1945)
*The Berenice Tiara*, 1899
aluminum, ivory, and garnets
cat. 2.39

Although French designer René Lalique worked with a variety of rich and rare materials throughout his career, often in combination, aluminum appears infrequently in his repertoire. His best known aluminum objects are a tiara made in 1899 and a series of cosmetics boxes and prototypes from around 1922.

Lalique created a special niche for himself as a maker of jewelry for use in theater productions. Stage jewelry must be oversized, large enough to be seen by an audience. Aluminum was an ideal material, since it relieved actresses of the burden of wearing heavy props. (Later, Hollywood would embrace aluminum for the movies, particularly when suits of armor were required.) The 1899 tiara was made for Comédie-Française actress Julia Bartet for her title role in Racine's tragedy *Berenice*. It consists of an aluminum frame made of a series of lotus flowers and openwork palmettes inset with five ivory plaques carved with scenes from the life of Berenice. Between each plaque is a figure of Isis, and both the lotus flower and palmette motifs are set with garnets.

In 1895, Lalique cut his ties with jewelry firms like Cartier and began creating exclusively under his own name. By 1901, he had developed a distinctive style of organic, naturalistic design which became synonymous with Art Nouveau. His materials palette expanded to include enamels, horn, ivory, and crystal glass. In the early 1920s, he again worked with aluminum, in an application that was to become highly significant for the material as the century progressed—packaging.

In 1906, Lalique was commissioned to produce pressed-glass perfume bottles for François Coty. This commission marked his shift from jewelry to glass, and also introduced him to the cosmetics world. A 1911 glass box produced for Coty proved too expensive to manufacture, so Lalique experimented with less costly materials such as aluminum. He made a series of prototype aluminum disks and boxes, often finished in a two-color lacquer process, but these were never manufactured. The only Lalique aluminum boxes to be produced commercially, for powder, were distributed in 1922 by Roger & Gallet.

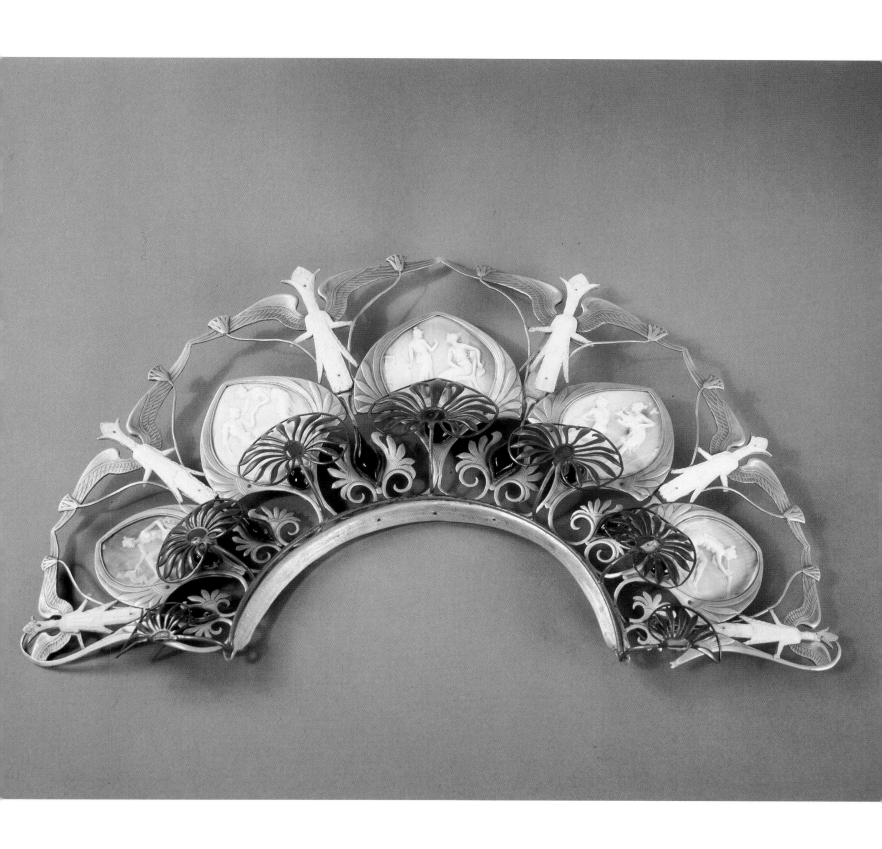

Dürener Metallwerke AG advertisement
*Aluminium* (January 1937) XXVII/1

Hugo Junkers (German, 1859–1935)
Suitcase, ca. 1919–20
duralumin
cat. 3.23

Construction of the Airship Akron
hull, 1929–31

Margaret Bourke-White
(American, 1904–1971)
Photograph of Airship Akron, set in
frame made from airship parts, 1931
duralumin and silver gelatin print
cat. 3.19

On July 2, 1900, Count Ferdinand von Zeppelin completed a test flight in his cigar-shaped airship made of an aluminum framework buoyed by seventeen hydrogen-gas balloons. The Zeppelin marked the beginning of a "new era of aerial locomotion," as reported by *The Aluminum World*. But the airship era really got underway with the development of duralumin, patented by the German metallurgist Alfred Wilm in 1910, a heat-treated aluminum-magnesium and copper alloy comparable to mild steel. When German airships were deployed in military operations during World War I, the Allies realized that these weapons could not be ignored. In 1916, the U.S. Navy approached Alcoa to develop a strong alloy for use in airships and thus the U.S. airship program was born. This also proved to be a turning point for Alcoa, since such a program required a systematic and directed approach to research and development.

Eventually, in the 1920s and early '30s, the U.S. Navy commissioned a number of airships such as the Shenandoah and the Akron, with frameworks made from a new aluminum alloy equivalent in strength to duralumin. Margaret Bourke-White photographed the Akron's unveiling, and presentation prints were framed with elements of the airship's girder by its builder, Goodyear. The interior fixtures and fittings aboard airships, particularly the luxurious passenger airships, were often made of aluminum as well, to reduce weight. The Hindenburg, for example, was equipped with aluminum furniture and a specially built aluminum piano. Passengers could even travel with light-weight aluminum luggage such as that produced by metal-airplane entrepreneur Hugo Junkers.

In the end, both the Shenandoah and the Akron crashed, as did the Hindenburg, putting an end to the deployment of airships as military craft and as a means of civilian transport. Aluminum remained aloft, however, as the fledgling aircraft industry developed and planes, not airships, began to rule the skies.

**Giuseppe Merosi,** designer
(Italian, 1872–1956)
Alfa Romeo, manufacturer
(Italian, 1910–present)
40/60 Aerodynamica, 1914
aluminum and other materials

Rolls-Royce (British, 1906–present)
Silver Ghost, 1907
aluminum and other materials

(opposite)
Bavarian Motor Works AG,
manufacturer
(German, 1916–present)
Wendler AG, manufacturer
(German, 1840–present)
BMW 328, 1938
aluminum and other materials

In September 1899, *The Aluminum World* reported that "horseless carriages are becoming much in evidence in large cities, and it is interesting to note that aluminum enters into their construction. The better class of motor vehicles have their gear case housings made of aluminum, and many manufacturers are experimenting and endeavoring to make more parts of the light metal." The fledgling automobile industry in both Europe and the United States made increasing use of aluminum, although primarily hidden under the hood, prompting the same publication to comment that "few people realize... the extent to which aluminum is used in the construction of automobiles." Perceptions began to change by 1907, with the polished aluminum body of the Rolls-Royce Silver Ghost.

Many automobile companies that are now household names were founded in the first two decades of the 20th century, among them Rolls-Royce, Alfa Romeo, and BMW. Metal paneling for cars probably originated in 1902 in Belgium with the Mercedes body (commissioned by King Leopold II), which established the practice of inserting aluminum panels into a wooden frame for high-quality bodywork. Rolls-Royce's reputation for excellence is based on the Silver Ghost, which combined the arts of the engineer and the coach builder. The company introduced styling changes infrequently and imperceptibly, and the Silver Ghost bodywork remained traditional in design throughout its eighteen-year history. The same cannot be said for the A.L.F.A. 40/60 Aerodynamica of 1914, with its sleek, cylindrical, tapered aluminum body and porthole windows. After World War I, the company—by then known as Alfa Romeo—produced a series of successful sports and touring cars, several with aluminum bodies.

Aluminum was used extensively in racing cars in the 1930s. Formula racing regulations limited the body weight of cars to 750 kilograms in 1930. There was no restriction, however, on the cubic capacity of engines, so lighter bodywork allowed larger and more powerful engines. A group of successful German racing cars of the period were dubbed the "silver arrows" because of their unpainted aluminum bodies. Between 1936 and 1940, BMW produced the 328, one of the most successful sports cars of its time. Three custom versions were made with a streamlined aluminum body; the reduction in weight increased their top speed by as much as 20 mph.

**Eileen Gray** (Irish, 1878–1976)
E.1027 cupboard, open, 1923–28
aluminum, wood, cork, glass,
and metal
cat. 3.36

(opposite)
**Eileen Gray** (Irish, 1878–1976)
E.1027 cupboard, closed, 1923–28
aluminum, wood, cork, glass,
and metal
cat. 3.36

Several French designers and architects in the late 1920s added aluminum to their repertoire of materials (Highlight 11). One such designer was Eileen Gray, now recognized as a pioneer of modern design. Born into an aristocratic Irish family, Gray trained in London before moving to Paris in 1902. Initially, she mastered the laborious Japanese lacquer-making technique, producing unique panels, screens, and furniture. Gradually she extended her professional interests to interior decoration. By the mid-1920s, she had moved away from such traditional, luxurious materials as lacquer to more modernist, industrial ones like aluminum. This shift signaled a completely different approach in her work, exemplified by the house she designed for herself and Jean Badovici at Rocquebrune in southern France.

Known as E.1027, the house showcased Gray's sensitive attention to detail and her ingenious and flexible brand of modernism. "From the richness of her lacquer pieces," writes biographer Peter Adam, "she had come to metal, glass, and plain wood without falling into the trap of mere functionalism." This is evident in the freestanding cupboard with an aluminum exterior designed for the main bedroom of E.1027. The cupboard served as a wall between the bedroom and a small washing area, and its aluminum back must have created some interesting reflective effects in the bedroom. When opened—using the aluminum scroll handle at the top—the cupboard's right-hand door created more privacy for the washing area.

Gray used aluminum for a number of furniture surfaces at E.1027, such as drawer fronts in the kitchen cabinet and the built-in bar in the dining room. Gray herself thought that aluminum was "an excellent material whose coolness is agreeable in a warm climate." It also provided visual and textural contrast to the other materials she used, such as cork. Many of the furniture pieces made for E.1027 served as prototypes from which she derived inexpensive versions in her Paris shop. She produced at least two other versions of the aluminum and cork room divider/cupboard. This object represents an innovative use of aluminum at a time when the metal was being embraced by leading Parisian furniture designers, although primarily for limited editions (figs. 37–38) and not as a material for mass production.

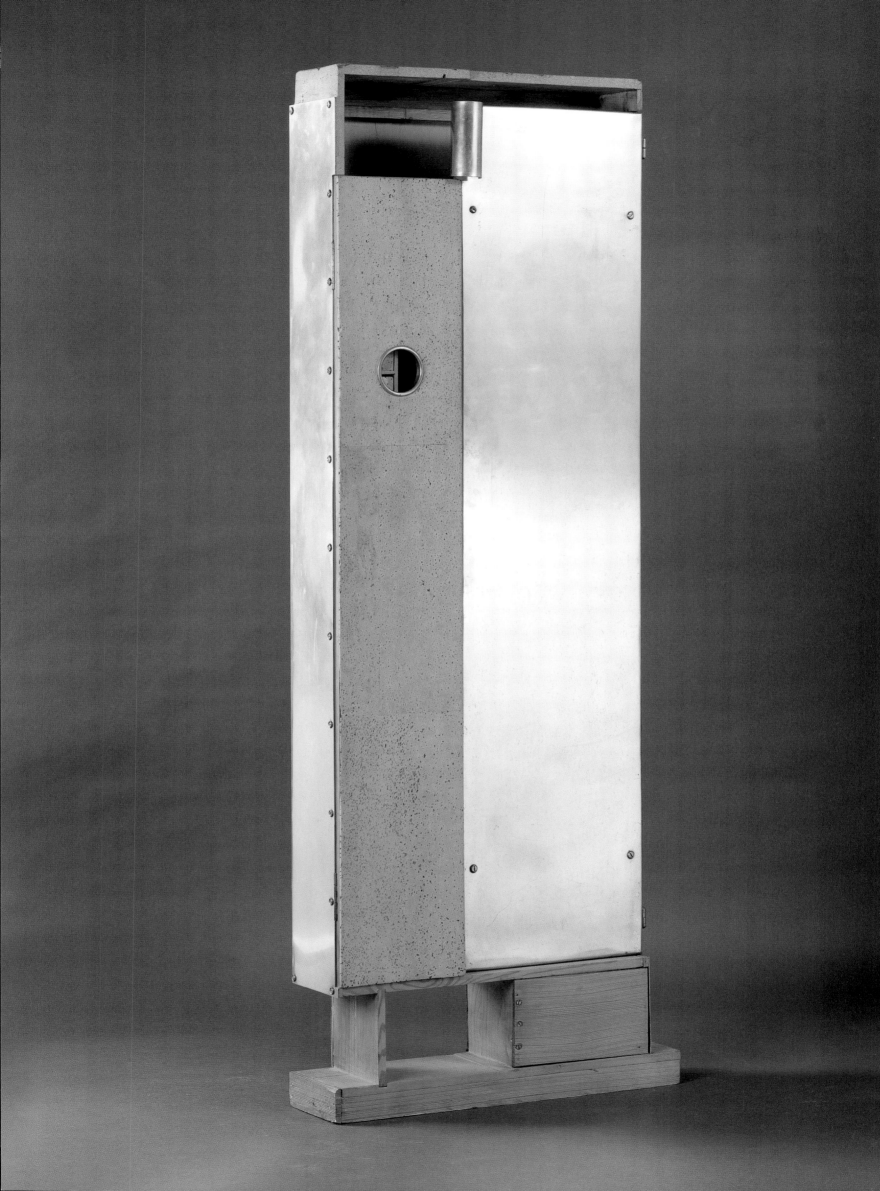

**Jacques Le Chevallier**
(French, 1896–1987)
Desk lamp, ca. 1927–30
aluminum and Bakelite
cat. 3.11

**Jacques Le Chevallier**
(French, 1896–1987)
Chistera lamp, ca. 1929
aluminum
cat. 3.12

**Jacques Le Chevallier**
(French, 1896–1987)
Desk lamp, ca. 1927–30
aluminum and Bakelite
cat. 3.13

(opposite)
**Jacques Le Chevallier**
(French, 1896–1987)
Desk lamp, ca. 1927–30
aluminum and Bakelite
cat. 3.10

Aluminum is not as reflective as silver, but the latter's price usually prevents its use in lighting for the everyday world. In the 19th century, aluminum trade publications noted the metal's advantages in lighting, particularly for reflector elements: a dull aluminum surface apparently threw a bright but soft light that was more evenly diffused than light reflected from a glazed surface. Most 19th-century uses of aluminum in lighting were nondomestic, such as acetylene bicycle and railroad lanterns and safety lamps for coal mines. But between the two world wars, the "material of the modern age" was used extensively in domestic lighting by designers on both sides of the Atlantic, including Jacques Le Chevallier, who produced an astonishing group of small aluminum table lamps between 1927 and 1930.

Le Chevallier was born in Paris and studied at the Ecole Nationale des Arts Décoratifs from 1911 to 1915. His interest in stained glass led him to master glassmaker Louis Barillet, with whom he worked after World War I. Barillet collaborated with noted French architects of the time, and his studio was part of the artistic hub of Paris. He and Le Chevallier were founding members of the Union des Artistes Modernes (UAM), along with designers and architects like Robert Mallet-Stevens, René Herbst, and Pierre Chareau.

Although lamp design represents a brief interlude in Le Chevallier's career as a stained-glass artist, his work in this area stretched the limits of materials and the mechanics of lighting, a contribution noted at the time in *L'Art Vivant* (although the author was uncertain whether his designs were driven by logic or the weird). Architectonic in form and inspired by Cubist principles, Le Chevallier's lamps are made from aluminum accented with other new materials such as Bakelite. The most powerful design is actually the simplest, Chistera, named after and based on the wicker basket used in the handball-like game jai alai.

American
Doors at Alcoa's Aluminum Research
Laboratories, New Kensington,
Pennsylvania, 1929

Wendell August Forge, manufacturer
(American, 1923–present)
Main gate at Alcoa's Aluminum
Research Laboratories, New Kensington,
Pennsylvania, 1929
aluminum
cat. 3.4

(opposite)
American
Overdoor decoration at Alcoa's
Aluminum Research Laboratories,
New Kensington, Pennsylvania, 1929
aluminum
cat. 3.3

Wendell August Forge, manufacturer
(American, 1923–present)
Main gate at Alcoa's Aluminum
Research Laboratories, New Kensington,
Pennsylvania, 1929
aluminum
cat. 3.4

In 1930, Alcoa opened a new research facility at New Kensington, near Pittsburgh. Not surprisingly, the building itself became a showcase for aluminum. The design by Henry Hornbostel incorporated aluminum in every conceivable way, both structurally and as ornamentation: it was used in the elevator doors, as inlay in the terrazzo floors, for window casings, piping, furniture, railings, and paint, and in cast ornamentation on the exterior, which was either polished, enameled (in blue, green, white, or red), or left with its natural sand-cast finish. Junius Edwards, assistant director of research, proudly observed that the laboratories were "not only a tribute to the versatility of aluminum and its alloys, when used for construction and laboratory furnishings, but represent many innovations in design and equipment." The wrought aluminum gates represent one such innovation.

Hornbostel had intended that the imposing entrance gates be made of the more typical cast aluminum. However, these plans changed when Wendell August, who operated a forge producing ornamental ironwork in Brockway, Pennsylvania, bid for the job. Although his craftsmen had no experience with aluminum, after much experimentation they mastered the material, and Wendell August Forge was awarded the contract for the gates. Given the research laboratory's mandate to develop new applications, the choice was highly appropriate as it paved the way for an expansion of aluminum into the market of hand-wrought architectural metalwork.

Wendell August Forge also produced the facility's elaborate hand-hammered repoussé aluminum elevator doors, generating yet another source of business. Hornbostel asked August to make some aluminum trays with repoussé designs based on the elevator doors. The trays were presented as gifts to Alcoa executives and eventually came to the attention of Pittsburgh department-store magnate Edgar Kaufmann, Sr., who commissioned the forge to develop a line of hand-wrought aluminum giftware. These proved to be a great success and August's business flourished. Others entered the market, including several artists who began their careers with Wendell August Forge, among them Arthur Armour (fig. 5). As a 1946 Alcoa advertisement declared, "the gate begat a tray; the tray begat a business... Since then, many others have gone into it and done well."

Warren McArthur (American, 1885–1961)
Armchair, 1932
lacquered steel tubing, lacquered wood,
and upholstery
cat. 3.42

Warren McArthur (American, 1885–1961)
Two-tiered table with two sample color
bars, ca. 1933

(opposite)
Warren McArthur (American, 1885–1961)
Rainbow Back chair, 1934–35
anodized aluminum and upholstery
cat. 3.43

Intrigued by the idea of making furniture with standardized parts, Warren McArthur produced his first metal furniture around 1924, using gas piping and car washers, a year before Marcel Breuer's tubular steel chair. By 1936, he had translated this novel approach into a furniture business generating nearly six hundred exclusive designs made from anodized aluminum. McArthur's combination of labor-intensive and self-evident construction methods with new materials and designs melded an Arts and Crafts sensibility with a celebration of the modern.

A mechanical engineer by training, McArthur was also an adventurous entrepreneur and an inventor with a great aptitude for problem solving. He moved to Arizona in 1913 and eventually, with his two brothers, became a partner in the Arizona Biltmore Hotel in Phoenix, for which he designed some furniture. Unfortunately, this project ran afoul of the Depression and the brothers lost control of the property. McArthur then moved to Los Angeles and, in 1930, formed his own furniture business. An early directory lists him as a custom furniture manufacturer working in steel, brass, and copper. He registered several patents at this time that defined his new construction methods: standard tubular parts joined by rings at their intersections, with an internal system of rods to provide the necessary strength and support.

McArthur initially produced tubular steel furniture, such as the gold-lacquered armchair illustrated here. By 1933, he was focusing exclusively on aluminum furniture with an internal skeleton of steel. In his choice of materials, he was no doubt influenced by the growing popularity of aluminum among designers and furniture makers. But there were practical considerations as well. Lighter than steel, aluminum was more suited to McArthur's construction methods. Its resistance to corrosion made it ideal for use on ocean liners, a lucrative market. And it could be colored with the anodizing process, in which the color becomes integral to the metal and thus will not chip or crack (although over time it does fade in light, clearly evident in the Rainbow Back chair). When McArthur's Los Angeles business went bankrupt, he reestablished the company in New York. With the advent of war, he switched to production of aluminum seating for the government, as his construction techniques and patents met the rigorous military specifications for stress, and aluminum was only available for military uses.

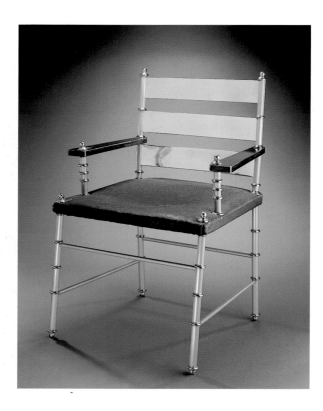

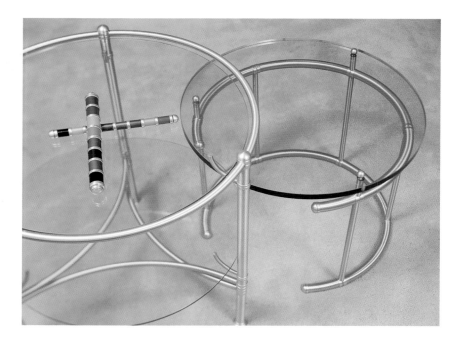

## 14. Marcel Breuer and the 1933 Aluminum Seating Competition

Photo of the Concours International du Meilleur Siège en Aluminium (International Competition for the Best Seating in Aluminum), Paris, 1933
*Revue de l'Aluminium* 60 (1934): 2384

**Herbert Bayer,** designer
(American, b. Austria, 1900–1985)
Wohnbedarf, retailer
(Swiss, 1931–present)
Cover of *das federnde Aluminium-Möbel* (Springy Aluminum Furniture), 1933
paper
cat. 4.13

**Herbert Bayer,** designer
(American, b. Austria, 1900–1985)
Wohnbedarf, retailer (Swiss, 1931–present)
Page from *das federnde Aluminium-Möbel* (Springy Aluminum Furniture), detail, 1933
paper
cat. 4.13

(opposite)
**Marcel Breuer,** designer
(American, b. Hungary, 1902–1981)
Embru-Werke AG, manufacturer
(Swiss, 1904–present)
Side chair, model no. 301, 1932
aluminum and bent plywood
cat. 3.37

In early 1933, the International Bureau for Applications of Aluminum, which had a vested interest in promoting wide-spread use of the metal, announced a competition for aluminum seating. Designs were to be economical, suitable for mass production for use in ships, offices, hotels, cafés, and theaters, and not simply imitations of furniture made in tubular steel or nickel. Participants from fourteen countries submitted 209 chair designs, 54 of which were realized in prototype. The accompanying exhibition featured several chairs by Alcoa, although these were excluded from the competition since they were already in production.

The competition brief stressed the importance of originality and of designing for aluminum's unique characteristics—especially its light weight. Few of the entrants complied with this, as was noted at the time in the *Revue de l'Aluminium*. One exception was Marcel Breuer. During his tenure at the Bauhaus, Breuer experimented with new materials such as tubular steel for furniture, advocating design based on a given material's innate properties. His drawings for a springy or pliant chair, patented in 1932–33, became the basis for various aluminum chairs produced by the Swiss company Embru-Werke and others from 1933 onward.

Breuer's competition entry included five prototypes: a stool, office chair, side chair, armchair, and chaise longue. Their solid-strap construction would have been impractically heavy in any metal but aluminum, as Breuer intended these chairs to be portable. He was awarded first prize by two independent juries, one composed of industry representatives, the other convened under the auspices of the International Congress of Modern Architecture. By 1934, Breuer's competition chairs were in full production, and had already received substantial press in both design and aluminum circles.

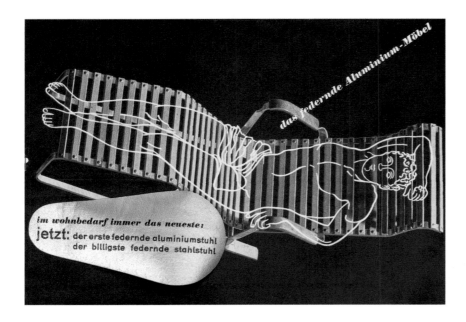

Alcoa advertisement
*Fortune* (September 1935): 36

Bohn Aluminum & Brass Corporation
advertisement
*Fortune* (June 1946): 53

**Lurelle Guild** (American, 1898–1986)
Drawing of "Streamlined Train"
for the "Aluminized America" show
at Marshall Field's department
store, Chicago, 1942

(opposite)
**Egmont C. Arens,** designer
(American, 1889–1966)
**Theodore Brookhart,** designer
(American, 1898–1942)
Hobart Manufacturing Co., manufacturer
(American, 1897–present)
Streamliner meat slicer,
model no. 410, designed 1940,
manufactured 1944–85
aluminum, steel, and rubber
cat. 3.95

Streamlining, the dominant design style of the 1930s and '40s in America, was synonymous with dynamism and modernity. Streamlining combined technology and design aesthetics in commercially innovative ways just as manufacturing companies were recovering from the Depression. Although streamlined objects were popular in the home and workplace—an example being the Hobart meat slicer—the design aesthetic found a niche in transportation because of its obvious association with aerodynamic movement. Aluminum was the perfect material for the streamlined age: clean, smooth, sleek, new, and readily cast or extruded into the seductive curves that defined the look. "In using extruded shapes," the *Aluminum News-Letter* reported in an article about streamlined trains, "the engineer designs the shapes he needs to fit the contour of the structure rather than designing the structure to fit conventional shapes." Most importantly, aluminum's lightness translated into greater speed: all-aluminum streamlined trains provided a smooth, comfortable ride *and* broke speed records.

As early as 1930, Alcoa advertisements were predicting that "one of these days you will ride on an Aluminum Train." In 1934, Union Pacific inaugurated the first all-aluminum streamlined train designed by E. E. Adams. That same year, Union Pacific's new model traveled from Los Angeles to New York City in a little under 57 hours, reaching a one-time peak record of 120 miles per hour. The achievement inspired the British government to invest in streamlined trains for the British railway system. In 1935, the New York, New Haven and Hartford Railroad introduced The Comet, an aerodynamic, bullet-shaped aluminum train that weighed 253,000 pounds fully equipped as opposed to the 700,000 pounds of a conventional steam train. Even trains built with steel as the chief structural metal still used aluminum for window frames, baggage racks, seat frames, and other fixtures and fittings to help reduce weight. By 1939, ten all-aluminum streamlined trains were in service across the United States, and two additional "up-to-the-minute aluminum flyers" were being built for the Missouri Pacific Railroad. World War II temporarily halted aluminum's invasion of the rails. At the end of the war, however, streamlined design still embodied the forward-looking optimism and promise of a new and better world, evident in the Bohn Aluminum & Brass Corporation advertisement illustrated here.

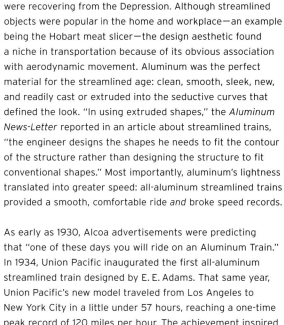

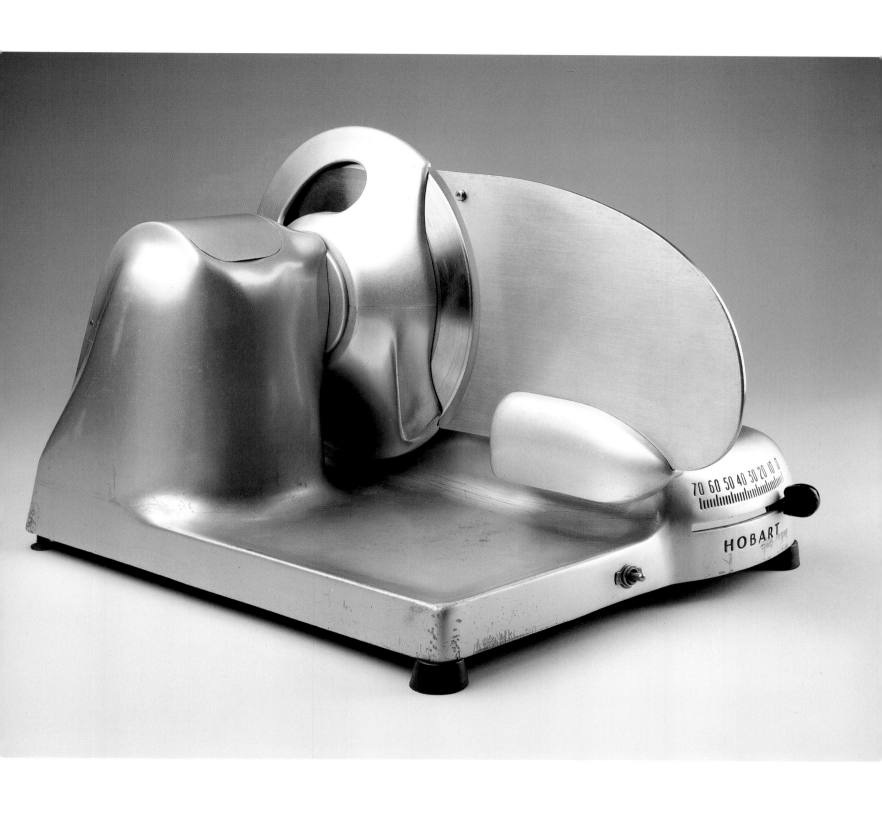

French bicycle racer Latourneau pulls
an Airstream trailer, 1947

The Shady Dell vintage trailer park,
Bisbee, Arizona, 1927–present

Airstream rally formation, Auburn,
Washington, 1962

(opposite)
Airstream factory: Demonstrating the
light-weight Airstream body, 1965

Mobility became a standard feature of American life during
the Great Depression, as millions of people, uprooted by
economic hardship, hit the road in search of employment.
Many worked on government-sponsored projects building
coast-to-coast highways, national parks, campgrounds,
and wilderness lodges. As the crisis eased and disposable
income increased, the stage was set for the travel-trailer
boom of the late 1930s.

Trailers allowed travelers to take the comforts of home
on the road. The aluminum Airstream Clipper went on the
market in 1936, the first trailer produced by Wally Byam's
newly established Airstream Incorporated. Eight hundred trailer
companies were operating in America at the time. Byam's
first trailer designs were issued in 1934 as blueprints, available
to the hobbyist to build at home. He was also a salesman for
the Bowlus-Teller Company, which sold aluminum trailers.
As a former designer for the aircraft industry, William Bowlus
was no stranger to aluminum. He realized its potential for
solving the problem of creating a strong, rigid trailer light
enough to be towed at speed by a mid-sized car. Unfortunately,
in 1935, overspending forced Bowlus-Teller into bankruptcy,
at which point Byam assumed control of the company and
renamed it Airstream Incorporated.

With its monocoque body of riveted aluminum, the Airstream
Clipper—named for the Clipper airplane—embodied the
streamlined and aerodynamic design of the time. To demonstrate the Airstream's lightness and maneuverability, in 1947,
Byam invited the famous French bicycle racer Latourneau to
visit the factory and pull a trailer with his bike. Today Airstream
trailers are still going strong; in 1999, the company made
*I.D. Magazine*'s design top-forty list for its contributions to
"aerodynamic living." But the Airstream trailer is also a collector's item and cultural icon, not only because of its practical,
streamlined design, but also because of the lifestyle and sense
of community that constitute the travel-trailer experience.
Through the Caravan Club International and its more recent
offshoot, Vintage Airstream Club, trailer devotees live out
Byam's motto, "adventure is where you find it, any place,
every place, except at home."

**Margaret Bourke-White**
(American, 1904–1971)
Untitled, ca. 1934

**Margaret Bourke-White**
(American, 1904–1971)
Alcoa commission photograph
("The bootlegger's enemy—
tamper-proof aluminum seals")
*Aluminum News-Letter* (March 1936): 3

**Margaret Bourke-White**
(American, 1904–1971)
Lobby of the Alcoa museum,
New York, 1935

(opposite)
**Margaret Bourke-White**
(American, 1904–1971)
Alcoa commission photograph
("An aluminum sheet a mile long, one
tenth as thick as this paper—that's foil!")
*Aluminum News-Letter* (April 1936): 3

Internationally renowned as a photojournalist who vividly captured the history-making events of her time, Margaret Bourke-White in her early career produced a significant body of corporate work, for either advertising or documentary purposes. As pointed out in the catalogue of a recent exhibition of her photographs for the International Paper and Power Company, "Bourke-White's photography has been inextricably intertwined with corporate culture," yet scholars have neglected her corporate publications in their studies. Certainly Bourke-White's relationship to Alcoa deserves further investigation. Beautifully composed images by Bourke-White documenting the production of aluminum—from the mining of bauxite to the making of foil and wire, focusing on production processes and the details of machinery or capturing the repetitious patterning of groups of objects—were published on a regular basis in Alcoa's *Aluminum News-Letter* from 1934 to the end of the decade, many as full-page presentation photographs.

Bourke-White's earliest photographs taken in Alcoa plants date from 1930, when she was working on assignment for *Fortune*. In September 1934, the magazine produced an article on Alcoa illustrated with her photographs. However, by that time she was also working independently for Alcoa, which had commissioned her to produce a photomural of aluminum production processes for the company's exhibit at the 1934 Century of Progress International Exposition in Chicago. Her bold, monumental industrial images lent themselves to the mural format. Alcoa again used Bourke-White's work for a large photomural installed in the lobby of the company's New York showroom, or "museum," designed by Lurelle Guild (Highlight 19) and opened in 1935. Total payment to Bourke-White for the showroom photos was $219.24, which broke down as follows: "prints for design $67.32, enlargements for lobby $145.80, tax on enlargements $6.12."

Russel Wright (American, 1904–1976)
Ravenware presentation drawing for
a pitcher and tumbler, ca. 1953
charcoal and chalk on colored paper

Russel Wright (American, 1904–1976)
Oval Rolly Cart, ca. 1932

(opposite)
Russel Wright (American, 1904–1976)
Lemonade pitcher and beverage set,
ca. 1932
spun aluminum and walnut
cat. 3.76 and 3.77

Russel Wright (American, 1904–1976)
Coffee urn, ca. 1935
spun aluminum and walnut
cat. 3.79

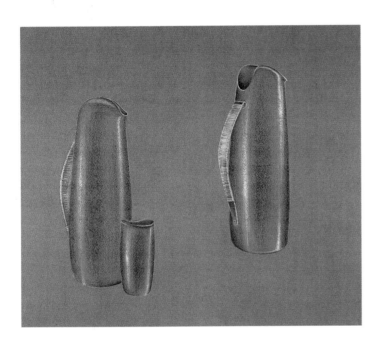

"One of the outstanding 'idea' men of the aluminum ware business," according to *House Furnishing Review* in 1933, Russel Wright brought aluminum out of the kitchen and into the center of informal entertaining at a time when American domestic lifestyles were changing dramatically. Although Wright favored the look of pewter and chromium-plated steel, aluminum was much easier to work, a critical difference since Wright—with his wife Mary Einstein—planned to manufacture the objects himself. "Aluminum, known to me only as a material for kitchen utensils, was beginning to be used in the infant aeronautics industry," Wright later explained. "Acquisition of a few tubes and sheets of aluminum was exciting for a young designer who wanted to design everything and only had a work bench and a spinning lathe. The easy workability of the metal, its permanent integral coloring inspired me to develop a variety of products…"

In 1930, Wright designed and produced his first informal serving ware made of spun aluminum. The designs reflect his preference for rounded, somewhat exaggerated forms, which were ideally suited to the spinning process. He often combined "cool" aluminum with "warm" materials such as wood, cork, and rattan to create contrasting textures and tones. His products were an immediate success and his shop became a destination for those interested in modern design. Alcoa also sent a delegation to examine the shop and collect information on sales. Shortly thereafter, Alcoa established the Kensington Ware line of aluminum serving and gift wares (fig. 141). Wright had uncovered a potentially vast market for aluminum that his fledgling business could not satisfy. Kensington Ware, on the other hand, backed by Alcoa's resources, could supply the market. Wright responded to this new competition by expanding his product lines, taking aluminum out of the house altogether and into the garden—on his bamboo "rolly cart." His press release on the lawn or porch picnic, "a time honored and typically American custom," included the photograph illustrated here. The objects on the cart, a punch-bowl set and wine cooler, are made of aluminum and walnut. Their light weight made "maidless and butler-less serving easy."

A versatile designer, Wright worked in a variety of media and for different companies. "In retrospect," he observed in a 1960 article for *Interior Design,* "it seems to me that aluminum and chromium were perhaps the most characteristic and the most popular materials used by modern designers until World War II. The white metal look gave a character and a mood to our creations that satisfied us." He regretted that the "white metal look" essentially disappeared after the war. Although he produced a set of designs in aluminum, wood, and rattan in the 1950s for Ravenware and SS-Sarna, these probably were never produced commercially.

**Lurelle Guild,** designer
(American, 1898–1986)
Aluminum Cooking Utensil Company,
later renamed Mirro Aluminum
Company, manufacturer (American,
1901–present)
Wear-Ever coffeepot, model
no. 5052, 1932
aluminum, wood, and paint
cat. 3.62

**Lurelle Guild** (American, 1898–1986)
Kensington Ware showroom,
New York, ca. 1938

(opposite)
**Lurelle Guild,** designer
(American, 1898–1986)
Electrolux L.L.C., manufacturer
(American, 1924–present)
Electrolux, model XXX, 1937
chrome-plated, polished, and enameled
steel, cast aluminum, vinyl, and rubber
cat. 3.59

The 1920s saw the establishment of an industrial design profession in America, and many companies turned to independent designers to modernize their products. This collaboration is exemplified by the career of Lurelle Guild. Beginning in 1927, Guild worked for many corporate clients, including Westinghouse, International Silver Company, Pitney Bowes, Norge Corporation, Montgomery Ward, Electrolux, and Alcoa. His design for the Electrolux model XXX vacuum cleaner resembled the new all-aluminum trains (Highlight 15), thus bringing the streamlined aesthetic and its association with speed and efficiency into the domestic environment. Objects designed for Alcoa's Kensington Ware and Wear-Ever lines—such as the Stratford compote (fig. 141) and the model 5052 coffeepot—epitomized the clean, strong forms popular between the wars and placed aluminum at the forefront of innovative design.

Guild's work with Alcoa went beyond a simple designer-client collaboration. In many respects, he became the design voice of the company in the 1930s. Alcoa's reliance on Guild is revealed in the amusing correspondence between the designer and Bill White of the Aluminum Cooking Utensil Company (an Alcoa subsidiary) concerning a costume design based on an aluminum coffeepot for a charity event in 1934, in which White states, "if you were not considered *the designer* for Kensington, you would not get this high grade assignment." In 1935, Guild designed the aluminum "museum" at Alcoa's New York offices at 230 Park Avenue (fig. 57). He produced drawings of buildings and interiors that were published in Alcoa's *Aluminum News-Letter* (fig. 102) and used in company advertisements. In 1942, he produced a series of stream-lined drawings for Alcoa featuring a train (Highlight 15), bus (fig. 31), airplane, washing machine, cooker, and weekend cottage, forecasting "the shape of aluminum things to come." These drawings, included in an exhibition at Marshall Field's department store in Chicago entitled "Aluminized America," were intended to pave the way for new postwar applications.

Guild continued to design for the company after the war, particularly for the Alcoa steamships, although he was not selected as a Forecast Program designer (Highlight 28), perhaps because by the mid-1950s Alcoa was looking for new talents. However, Guild was a prominent figure in product design who deserves greater recognition, particularly for his role in modernizing aluminum.

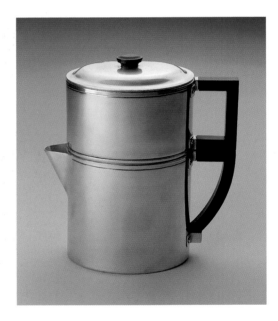

**Pierre Alexandre Morlon,** engraver
(French, 1878–1951)
Medal, recto and verso, 1937
aluminum
cat. 4.4

L'Aluminium Français, publisher
(French, 1911–83)
Page from *L'Aluminium à l'Exposition,
Paris 1937* (visitors' guide to aluminum
at the 1937 Paris Exposition), 1937
paper
cat. 4.5

(opposite)
L'Aluminium Français, publisher
(French, 1911–83)
Cover of *L'Aluminium à l'Exposition,
Paris 1937* (visitors' guide to aluminum
at the 1937 Paris Exposition), 1937
paper
cat. 4.5

International exhibitions have served as ideal promotional vehicles for aluminum producers and manufacturers since aluminum was first shown at the 1855 Paris Exposition, particularly as these venues showcased innovative advances in science, technology, design, and manufacturing. The 1937 International Exposition of the Arts and Techniques Applied to Modern Life, which covered 250 acres along both sides of the Seine, was the most ambitious effort to date to illustrate the value and importance of aluminum in modern architectural, industrial, and domestic applications. Aluminum was featured in many of the individual country and theme pavilions, and played a significant role in the construction of exhibition buildings, bridges, decorative panels, sculpture, and furnishings.

L'Aluminium Français, founded in 1911 to promote French aluminum producers, sponsored the Aluminum Pavilion, which consisted of a cinema presenting documentary films on aluminum and an exhibition gallery showing different applications of the metal. The organization also produced a booklet containing a map pinpointing the location of aluminum in all its manifestations—objects, architectural fittings, structural components—throughout the exhibition. A commemorative medal, issued by L'Aluminium Français, featured the pavilion facade on one side and listed the applications of aluminum on the other.

According to the general regulations, the exhibition was intended to unite "original works of craftsmen, artists, and manufacturers,… [to] be creative, educative, and even to bring forth designs which at the present time seem to belong to the future." However, the exhibition is remembered principally for the political ideology expressed in a number of the national pavilions. A prime example was the German pavilion designed by Albert Speer, in which aluminum carried ideological freight as a "German metal." German production of aluminum would outstrip that of the United States by 1938, making the country the largest producer in the world. German supremacy was announced in the pavilion's monumental entrance doors made of aluminum with gold and red anodized grille panels. The pavilion showcased an aluminum Mercedes-Benz racing car, which had established a new speed record in 1936. Although the Paris Exposition brought aluminum to center stage as a new material ideally suited to modern life, the strategic and political role it was poised to play was certainly an underlying theme.

# L'ALUMINIUM
## A L'EXPOSITION

*Paris*
1937

Cast spoons, still uncut, at the Swiss National Exhibition, Zurich, 1939

Aluminum Pavilion, Swiss National Exhibition, Zurich, 1939

(opposite)
**Hans Coray,** designer (Swiss, 1906–1991)
Blattmann Metallwarenfabrik AG, manufacturer (Swiss, 1838–present)
Landi chairs, pre-1962; designed 1938, manufactured 1939–present
molded, heat-treated, and stained aluminum and rubber
cat. 4.11

By 1938, Switzerland ranked sixth in the world in aluminum production, after Germany, the United States, Canada, France, and the Soviet Union. With an annual output of 28,000 tons, aluminum constituted 5–6 percent of total Swiss exports, and the metal's significance for the country was reflected in its representation at the National Exhibition held in Zurich between May and October 1939. The vast, rectangular edifice of the Aluminum Pavilion—a veritable exhibition within an exhibition—covered 3,000 square meters and contained over 160 exhibitors, including three foil manufacturers, eight cooking-pot and utensil companies, and two piston makers. Moreover, aluminum figured prominently in the construction of buildings and in different displays. For example, the Army Pavilion included aluminum pontoons and the Transport Pavilion featured aluminum railway cars and cable-car cabins. Many black-and-white photographs from the exhibition show aluminum in various stages of fabrication, from an array of extrusions (frontispiece) to an arrangement of aluminum spoons. These carefully composed images match the abstract power of Margaret Bourke-White's photographs for Alcoa (Highlight 17).

Today, the Swiss National Exhibition—or "Landi" (from Schweizerische Landesausstellung)—is remembered principally for the chair chosen via competition as its official seating. Hans Coray, a self-taught designer, submitted an innovative design that resulted in the Landi chair. Its continuous seat and back was formed with a perforated aluminum shell, an idea borrowed from the aviation industry. Perforation reduced the chair's weight and permitted rainwater to run off the seat, features essential for outdoor use. The chair was used again in the Swiss Pavilion at the 1958 World's Fair in Brussels. An icon of aluminum—and Swiss—design, the Landi chair has been produced continuously since 1939, although with some design modifications.

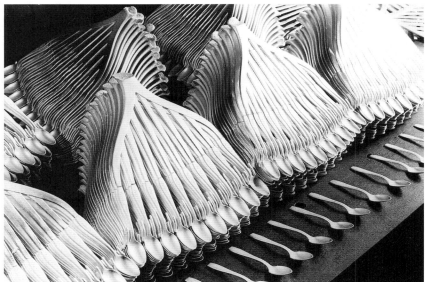

(left) **Gio Ponti** (Italian, 1891–1979)
Swiveling chair #4, prototype, from the
Montecatini Corporation headquarters,
Milan, ca. 1937
aluminum
cat. 4.16

(right) **Gio Ponti,** designer
(Italian, 1891–1979)
Ditta Parma Antonio e Figli,
manufacturer (Italian, 1870–present)
Swiveling chair #3 from the Montecatini
Corporation headquarters, Milan, 1938
aluminum, steel, and vinyl
cat. 4.17

Italian
Postcard, ca. 1927
aluminum
cat. 4.18

(opposite)
**Gio Ponti,** designer (Italian, 1891–1979)
Kardex Italiano, manufacturer
(Italian, active first half of 20th century)
Chair from the Montecatini Corporation
headquarters, Milan, 1938
aluminum, painted steel, and
padded leatherette
cat. 4.15

The grand master of Italian modernism, Gio Ponti excelled
in many fields: as an architect, industrial designer, inspired
teacher, writer, and editor. He opened an architectural studio
in Milan in 1921, after completing studies at the Milan Poly-
technic. In 1936, he was commissioned to design the Milan
headquarters of the Montecatini Corporation, Italy's largest
mineral and chemical company and foremost producer of
aluminum. Ponti designed the building along with all its fixtures, fittings, and furniture, including several different chair
designs. Judging from photographs taken at the building's
inauguration on October 28, 1938, these designs were intended
to correspond to different spaces, functions, and hierarchies.
The chair designs demonstrate Ponti's willingness to challenge
received notions of mass production by introducing variety
and combining standard parts in different ways.

The Montecatini building contained 350 tons of aluminum,
primarily in the roof, window frames, gates, balustrades, and
elevator doors. The office desks, chairs, and other furnishings were also made of aluminum. This was hardly surprising
given Montecatini's preeminence in the aluminum business,
but by this date aluminum's use in Italy also had nationalist
and political overtones.

In 1935, under Mussolini and the Fascist regime, Italy invaded
Ethiopia. The League of Nations punished this act with
economic sanctions. A few years earlier, the Fascists had introduced the policy of autarky, or self-sufficiency, whereby
all the country's indigenous resources—including aluminum—
were to be fully exploited and used in preference to imported
materials. The postcard made of aluminum, showing the
Fascist symbol of the fasces—a bundle of reeds bound together
with an axe—as a lamp throwing light on the exploits of the
Italian aviator Francesco de Pinedo, is an early example of
autarky. Under the economic sanctions imposed by the League
of Nations, this policy became a necessity, and aluminum the
material of political expediency. In this context, the Montecatini
building became a virtual manifesto of autarky.

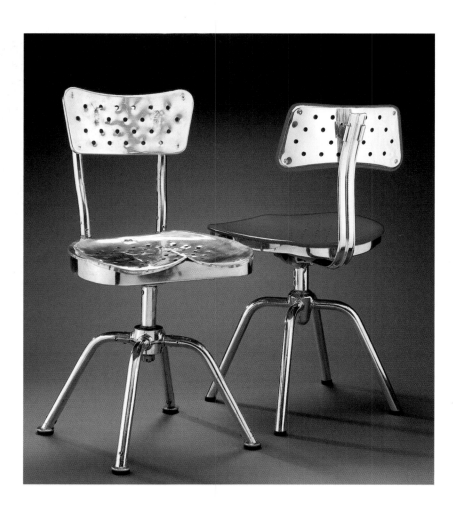

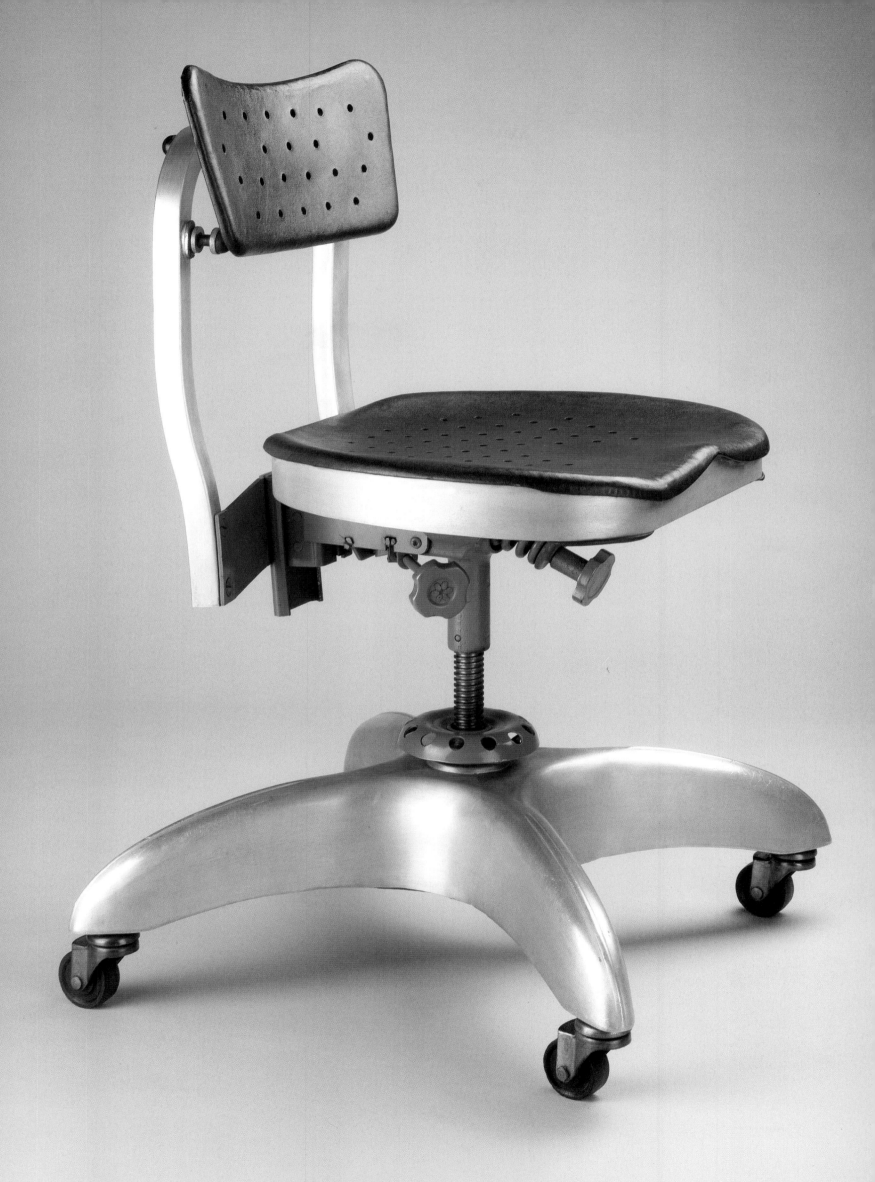

**Margaret Bourke-White**
(American, 1904–1971)
Men working on the construction
of Douglas DC-3 wings
*Aluminum News-Letter*
(April 1940): 3

**Margaret Bourke-White**
(American, 1904–1971)
Inspection of aluminum propeller blades
*Aluminum News-Letter*
(November 1939): 3

(opposite)
**Margaret Bourke-White**
(American, 1904–1971)
Sikorsky airplane, ca. 1934

All-metal planes were the norm by 1936, the year the *Aluminum News-Letter* proclaimed that "aluminum has become the *speed metal* of a new and faster age." The shift from wood to metal in airplane construction entailed years of research, which in the United States was largely supported by the military through funding and technical direction. Moreover, the decision to replace wood and fabric with aluminum was neither an obvious nor a straightforward one but, as explained by Eric Schatzberg in his book *Wings of Wood, Wings of Metal*, involved both technical and cultural criteria. The latter were shaped by the rhetoric promoting aluminum as the metal of the future with its overt connections to science, engineering, and metallurgy. The wooden plane, on the other hand, was seen—certainly by the mid-1930s—as having "the appearance of a product of the furniture trade."

Yet even early planes contained some aluminum components. For instance, Orville Wright's plane, first flown in 1903, carried an aluminum motor block. Aluminum's predominance in the field of aircraft construction began with developments in heat-treated alloys such as duralumin, which combined lightness with strength, and the advanced aeronautics research involving all-metal planes undertaken in Germany during World War I, particularly the innovative work of Hugo Junkers. After the war, Junkers turned to commercial air transport and in 1919 unveiled the F13, a five-passenger plane constructed entirely of duralumin. Ultimately, however, duralumin posed serious problems for aircraft designers: over time, duralumin sheet becomes brittle and eventually degenerates.

The benchmark for aluminum airframe construction was set in the late 1920s through experimental research into new aluminum alloys subsidized by the U.S. military. Through these efforts, the metal airliner came of age. Among the earliest was the Boeing 247, a twin-engine monoplane with an aluminum alloy structure, built in 1931. New models quickly followed, culminating in 1934 with the Douglas DC-3, "one of commercial aviation's most durable workhorses," built in the thousands during World War II. Today, aircraft construction continues to depend on versions of the aluminum alloys developed during the interwar period.

Swallow Sidecars Ltd., later renamed
Jaguar Cars (British, 1921 – present)
Gloster Meteor jet center section,
ca. 1944

Rolls-Royce (British, 1906 – present)
Array of jet engine parts in hangar,
ca. 1994

(opposite)
Rolls-Royce (British, 1906 – present)
Jet engine on a British Airways
Boeing 747-400, ca. 1989

Warfare has traditionally been a great stimulus of techno-logical innovations that are later adapted to peacetime purposes. The development of the jet engine during World War II set the stage for a revolution in postwar commercial air travel. Sir Frank Whittle is credited with obtaining the first patent on the jet engine in 1930. But the first operational jet engine was designed in Germany by Hans Pabst von Ohain and powered the first jet-aircraft flight in 1939. By 1944, the Royal Air Force's Gloster Meteor jet aircraft were battling it out with Germany's jet-powered Messerschmitt Me-262s. The Gloster Meteor engine was made by Rolls-Royce and its center sections were fabricated at Jaguar's Coventry plant.

Immediately after the war, commercial airlines continued to use the more economical propeller plane. But several compa-nies on both sides of the Atlantic realized that the future of air travel lay with the jet engine. The engine's efficiency was increased, and gradually aircraft makers, aided by develop-ments in aluminum alloys, designed planes especially for the jet engine and its superior capabilities. The Boeing 707 went into service in 1958, the first successful commercial jet. The new jets revolutionized the speed of air travel and reduced its cost, as larger planes could carry more passengers. In 1970, the Boeing 747, the first wide-bodied jetliner, introduced many new materials specifically developed for the aircraft. An aluminum alloy used in the wing, for example, reduced the plane's weight by 6,000 pounds. It also meant fast, relatively low-cost transportation on a scale never before available. As the president of Pan American stated, "the new era of mass travel between nations may well prove more significant to human destiny than the atom bomb. The 747 will be a great weapon for peace."

The aircraft industry provides a good example of how "material wars" advance technology. In the 1980s, aluminum accounted for nearly 80 percent of the weight of commer-cial aircraft, but developments in other materials such as "superalloys," titanium alloys, and composites threatened aluminum's preeminence. The aluminum industry responded by developing new alloys, for example, combining the metal with lithium. A key factor in the advancement of materials is the growing ability to "design" them with specific properties for specific needs, which has been crucial in the aeronau-tics industry.

R. Buckminster Fuller
(American, 1895–1993)
Construction of Wichita House,
Wichita, Kansas, 1947

R. Buckminster Fuller
(American, 1895–1993)
Construction of Wichita House,
Wichita, Kansas, 1947

R. Buckminster Fuller
(American, 1895–1993)
Wichita House, Wichita, Kansas, 1947

(opposite)
R. Buckminster Fuller
(American, 1895–1993)
Construction of Kaiser Dome,
Honolulu, 1957

R. Buckminster Fuller
(American, 1895–1993)
Grand-opening evening symphony
concert, Kaiser Dome, Honolulu, 1957

A romantic visionary who rejected preconceived notions of architecture and engineering, Fuller created utopian forms of housing with the tools of the scientist and philosopher. Technological efficiency and conservation of resources were axiomatic for Fuller, and "lightness" was one strategy for achieving both. Thus, aluminum played a key role in his futuristic vision.

Fuller's Dymaxion House (*dymaxion* combines *dynamic*, *maximum*, and *ion*), published as a model in 1929, introduced to the public some of his key concepts, including the industrially produced house priced by the pound. Between 1941 and 1946, Fuller modified his original design to incorporate newly developed light aluminum alloys with increased strength-to-weight ratios. At war's end, conditions were ripe for mass production of Fuller's design: vast supplies of aluminum and idled aircraft workers could be deployed to meet the urgent need for low-cost housing. Fuller worked with Beech Aircraft Company of Wichita, Kansas, who intended to build 60,000 Dymaxion Houses a year through the newly formed Fuller Houses Inc. Unfortunately, disagreements among the partners killed the project by 1947, and only prototypes were built. The Dymaxion Wichita House was made of a circular aluminum structure suspended from a central mast. Assembled by six workers in one day, it weighed 1.5 tons, as compared to the 150 tons of a conventional house.

The geodesic dome, Fuller's trademark and the only invention that brought him financial success, was patented in 1954. Although Fuller built geodesic domes of paper, plastic, and bamboo, the favored material was aluminum. The typical geodesic dome is composed of interlocking triangular elements made of extruded aluminum, then covered with a skin of sheet aluminum. Unlike classic domes, the geodesic does not depend on heavy vaults or flying buttresses. Rather, the structure is supported by tension between the struts. "Geodesic domes," according to J. Baldwin, editor of the *Whole Earth Review*, "get stronger, lighter, and cheaper per unit of volume as their size increases—just the opposite of conventional building." Thus, the geodesic dome fulfilled Fuller's vision of maximum efficiency and more-with-less, while allowing him to break out of the mold of "rectangular" architecture.

Industrial magnate and aluminum producer Henry Kaiser was among the many licensees of the geodesic dome and, in 1957, commissioned Fuller to design one as an auditorium in Honolulu. Once all the component parts arrived, the auditorium—fifty meters in diameter and made from aluminum panels and struts—was built by thirty-eight men in only twenty working hours.

Ernest Race, designer
(British, 1913–1964)
Race Furniture Ltd., manufacturer
(British, 1945–present)
BA3 Chair, designed 1945, manufactured
1945–69 and 1989–present
stove-enameled cast aluminum
and upholstery
cat. 4.20

(opposite)
Clive Latimer, designer
(British, dates unknown)
Heal & Son, manufacturer
(British, active 1940s)
Plymet cabinet, prototype, 1945–46
cast and sheet aluminum, sheet steel,
and birch veneer
cat. 4.19

A national exhibition held at the Victoria and Albert Museum in 1946, "Britain Can Make It" aimed to promote public awareness of design in British industry. The government, an initial force behind the exhibition, recognized that design could play a critical role in stimulating sales at home and abroad. The exhibition occupied 90,000 square feet of gallery space that had been emptied during the war, and displayed a multitude of products. One of its chief mandates was to explore the conversion of wartime technology to peacetime production.

Although "Britain Can Make It" included aluminum furniture, it was presented more out of necessity than desire. Ernest Race, an innovative designer and manufacturer of the aluminum BA3 Chair, himself believed that for furniture, "wood remains supreme, metal [being] an excellent alternative in times of shortage." But a postwar glut of aluminum, the need to employ thousands of workers skilled in light alloys fabrication, and severe shortages of traditional materials like wood combined to make aluminum an obvious choice for furniture. Race's BA3 Chair, shown at the exhibition, was made of aluminum resmelted from wartime materials. The chair's parts were originally sand-cast, but a foundry owner approached Race with a proposal to make the frame using a technique—pressure die-casting—developed during the war for incendiary bomb casings. The result was a more refined, thinned-down chair, its weight reduced by 25 percent. Popular in both institutional and domestic markets, the chair was also suitable for export since it could be shipped as easy-to-assemble parts.

"Britain Can Make It" was intended to showcase the country's industrial potential as well as actual capability, and manufacturers were encouraged to submit prototypes. Among these was Heal & Son's Plymet furniture line. The company borrowed a technique developed for aircraft construction of bonding wood to aluminum, thereby minimizing the amount of wood used but creating a woodlike effect. The Plymet sideboard had wood veneer only on the body; the doors and drawer fronts were made of bare aluminum at a time when most British furniture manufacturers tried to disguise the metal. One visitor to the exhibition thought it looked "too much like a fridge." The Plymet line never went into production, possibly because it was too advanced for the average Briton's taste at the time.

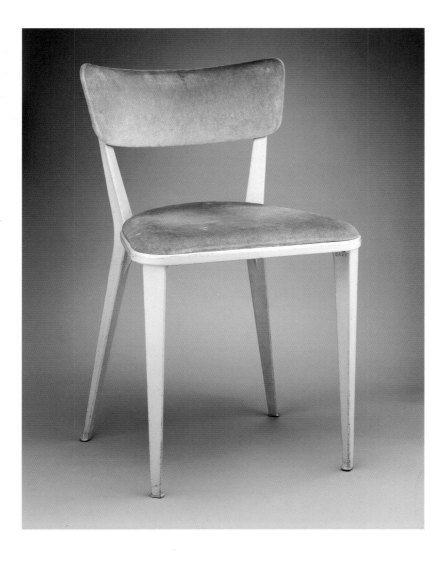

**Jean Prouvé** (French, 1901–1984)
Headquarters of the Fédération du
Bâtiment (Building Trades Federation),
facade, Paris, 1949

**Jean Prouvé** (French, 1901–1984)
Hoisting the first panel of the
Fédération du Bâtiment (Building
Trades Federation), Paris, 1949

**Jean Prouvé** (French, 1901–1984)
Wall light, 1951–52
aluminum and glass
cat. 3.18

(opposite)
**Charlotte Perriand,** designer
(French, 1903–1999)
**Jean Prouvé,** designer
(French, 1901–1984)
Bibliothèque Mexique, 1952
aluminum, steel, and wood
cat. 3.49

Working individually and in partnership, Charlotte Perriand and Jean Prouvé helped to shape the course of 20th-century French design and architecture. In 1952, they collaborated on the furniture for the Cité Universitaire in Paris. One design, the Bibliothèque Mexique, combined wooden horizontal elements separated by painted aluminum panels and diamond-pointed aluminum doors. This mix of wood and aluminum structural components that could be combined in many different ways formed the basis of numerous other bookshelves produced by Perriand and Prouvé.

Both designers worked extensively in metal from the beginning of their careers and, by the 1950s, both were familiar with aluminum. Perriand, who trained as a designer, embraced metal as the symbol of revolutionary design. Prouvé, who apprenticed as a metalworker producing wrought iron, brought to architecture and interiors a commitment to industrialization and faith in the benefits of metal. Perriand's first major project was the Bar under the Roof, a modern interior of steel and aluminum shown at the 1927 Salon d'Automne. In 1996, recalling that moment in her career, Perriand remarked, "I was absolutely in the epoch, I wore a big necklace of aluminum marbles." This acclaimed project led to work with Le Corbusier designing metal furniture. In 1939–40, she and Prouvé developed prefabricated, multipurpose aluminum cabins that were used as offices, a social club, and accommodations for an aluminum-alloy business. This was the springboard for Prouvé's postwar work with aluminum.

Prior to 1940, Prouvé had used aluminum primarily for minor architectural fittings such as door handles and railings. Following the war, however, it became an integral part of his work, perhaps because in 1949, the promotional association L'Aluminium Français—which wanted to expand the use of aluminum in architecture—purchased a 17 percent interest in his workshop. For the headquarters of the Building Trades Federation (1949–50), he introduced the aluminum curtain-wall technique. The facade was made of large, prefabricated aluminum panels that were hoisted into place. Prouvé continued to use aluminum panels for walls and doors, and designed aluminum blinds, shutter systems, light fixtures, and furniture combining aluminum elements with other materials. Aluminum is found in his residential commissions, his public buildings, and his work in France's African colonies, often incorporating his signature portholes and ribbed or diamond-pointed surfaces. L'Aluminium Français had placed its finances and faith in the right man.

Alcoa Forecast Program advertisement of Suzy Parker wearing Jean Desses ball gown as photographed by Richard Avedon (American, b. 1923)
*Time* 68 (November 26, 1956): 52–53

**Jean Desses**
(Greek, b. Egypt, 1904–1970)
Ball gown with stole, detail, 1956
Lurex, silk, and mink
cat. 4.26

(opposite)
**Jean Desses**
(Greek, b. Egypt, 1904–1970)
Ball gown with stole, 1956
Lurex, silk, and mink
cat. 4.26

By the early 1950s, aluminum supply had caught up with demand, and the three major American producers—Alcoa, Reynolds, and Kaiser—were attempting to develop new product areas and increase their share in traditional markets. For the first time, but in different ways, all three directed resources toward design as a vehicle for new ideas. Through design, the companies hoped to increase public awareness of aluminum and, of course, expand sales.

Alcoa's Forecast Program, primarily an advertising campaign designed to let the public "glimpse the lightness and brightness and beauty of aluminum that will come into your home and into your life," commissioned well-known designers to create objects made of aluminum but not typically associated with the metal. Forecast wanted unusual items, rather than objects for general production. Each object would be photographed for a nationally distributed magazine advertisement by a noted photographer, who would receive credit, along with the designer, in the ad. The first ad, which appeared in October 1956 in *U.S. News and World Report*, *Time*, *Newsweek*, the *New Yorker*, and the *Saturday Evening Post*, introduced the Forecast logo and its slogan, "there's a world of aluminum in the wonderful world of tomorrow."

The second ad, run as a double-page spread in the same magazines in November and December 1956, aligned the metal with haute couture and a cosmopolitan sensibility. It featured an "aluminum" ball gown designed by Jean Desses and worn by Suzy Parker, described as "probably the world's best-known model." Richard Avedon photographed Parker as she posed in front of the Eiffel Tower and "the elegant Paris restaurant Marigny."

Desses' gown was made from twenty-eight yards of woven Lurex (aluminum foil coated on both sides with plastic), intended to have the look and draping quality of chiffon. The gown toured extensively before it was given to the Brooklyn Museum's fashion collection in January 1957. Over the next couple of years, the price of aluminum yarn fell from about $17 to $6 a yard, resulting in its widespread use in clothing, particularly bathing suits and cocktail dresses.

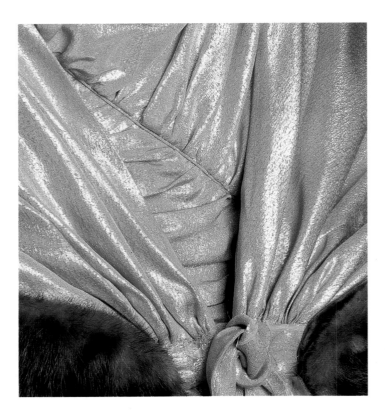

Cover of *Shelter* 2, no. 5
(November 1932) with Isamu Noguchi's
*Miss Expanding Universe*

Alcoa Forecast Program advertisement
with Isamu Noguchi's table
as photographed by Irving Penn
(American, b. 1917)
*Time* 69 (April 29, 1957): 48

**Isamu Noguchi**
(American, b. Japan, 1904–1988)
Drawings for "skin stress furniture," 1954
Ink on paper

(opposite)
**Isamu Noguchi**
(American, b. Japan, 1904–1988)
Alcoa Forecast Program tables, 1957
aluminum and paint
cat. 4.28

Isamu Noguchi embraced new industrial materials through-out a long and productive career, which regularly crossed and blurred the boundaries between fine and decorative art. He first used aluminum in sculpture in a 1932 work entitled *Miss Expanding Universe*. This piece, which reinforced aluminum's status as a modern and innovative material, evolved out of Noguchi's close friendship with R. Buckminster Fuller (Highlight 25). Fuller and Noguchi were part of a dynamic, avant-garde circle that embraced astronomer Edwin Hubble's theories of an expanding universe, which later would become the basis of modern cosmology. Noguchi's sculpture, a flying female figure with outspread arms and legs, reflects the new view of a seemingly limitless universe. The work was featured on the cover of Fuller's magazine *Shelter* in November 1932. The issue included Fuller's essay "Streamlining," which presented preliminary studies for his Dymaxion Car.

Noguchi's next major immersion in aluminum occurred in the late 1950s, when he used sheet aluminum supplied by Alcoa in his sculptures because he "wanted to deny weight and substance." Noguchi had previously worked with the company as a designer on its Forecast Program (Highlight 28), creating a three-part hexagonal table that was photographed by Irving Penn and advertised in April and May 1957. The Alcoa commission seems to have offered an opportunity to pro-duce an existing design, since a drawing of a similar table is inscribed "invented by Isamu Noguchi during October 1954." Although there is no indication on the drawing that the "skin stress furniture" represented was to be made of aluminum, it was certainly a practical choice. The Forecast tables were intended to be modular and either painted or anodized. As the advertisement stated, "you will style rooms to the whim of the moment... using sectional aluminum furni-ture of myriad textures, colors, finishes and forms... in arrangements as endless as the patterns of a kaleidoscope." Alcoa estimated that the tables could be "made for as little as $13." Unfortunately, the prototypes made for the Forecast Program never went into commercial production.

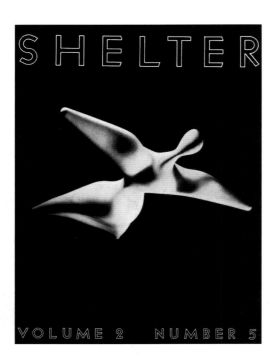

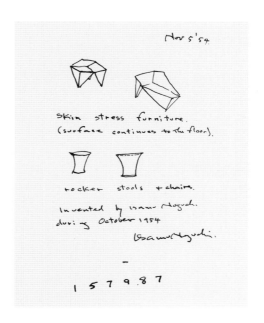

**Charles Eames,** designer
(American, 1907–1978)
**Ray Eames,** designer
(American, 1912–1988)
with the University of California
at Los Angeles
Side chair, full-scale model
of prototype, 1948
Painted neoprene-coated aluminum

**Charles Eames,** designer
(American, 1907–1978)
**Ray Eames,** designer
(American, 1912–1988)
*Solar Do-Nothing Machine*, 1957
anodized aluminum

(opposite)
**Charles Eames,** designer
(American, 1907–1978)
**Ray Eames,** designer
(American, 1912–1988)
Herman Miller, Inc., manufacturer
(American, 1923–present)
Aluminum Group armchairs,
with and without upholstery,
ca. 1965; designed 1958,
manufactured 1958–present
aluminum and vinyl
cat. 4.45

In 1948, the Museum of Modern Art in New York sponsored an international competition for low-cost furniture. The Eames Design Office and engineers from the University of California at Los Angeles shared second prize for their jointly designed seating prototype. Their entry featured a chair with a stamped aluminum shell seat and back, and cast aluminum pedestal base. Although the metal stamping process was not viable economically and the chair never went into production, aluminum had entered the material vocabulary of the Eames workshop.

The cast aluminum pedestal chair base was used extensively by Eames. Although this type of support structure is found prior to World War II, for example in the office furniture supplied by Alcoa to Mellon Bank in the early 1920s, it became ubiquitous following its use by Eames in the Aluminum Group, introduced by Herman Miller in 1958. The Aluminum Group, intended for indoor and outdoor domestic or institutional use, is still in production today. Along with an aluminum base, the chairs consist of a cast aluminum frame formed of two grooved side elements creating a continuous seat and back, with upholstery fixed tautly between two cross supports. The organic, flowing frame adds visual interest to the back and sides of the chair.

Simultaneous with the Aluminum Group commission, the Eames workshop was invited to participate in Alcoa's Forecast Program (Highlight 28). Given free rein, Eames produced the *Solar Do-Nothing Machine*, a moving toy consisting of flat circles, snowflakes, stars, and pinwheels of aluminum powered by sunlight. Eames felt that to give the object an instrumental purpose would obscure its "subtext": conservation and the economy of natural resources. Charles and Ray Eames were visionary designers and, if nothing else, this machine attests to their prescience. Aluminum and its relationship to energy expenditure and reclamation through recycling are now key issues in design, manufacturing, and consumption (Highlight 39).

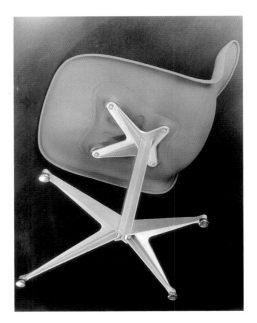

**Gunnar Larsen**, photographer
(Danish, 1930–1990)
Maud Betelsen in a dress by Paco
Rabanne sitting on a bench in front of
the church of St. Germain des Prés,
Paris, ca. 1968

Lamex, manufacturer (French, active
second half of 20th century)
Butcher's apron, detail, ca. 1970–80
aluminum and linen
cat. 5.4

(opposite)
**Paco Rabanne** (Spanish, b. 1934)
Atomium Bruxelles Dress, 1999
aluminum and stainless steel
cat. 5.2

Aluminum, plastic, Plexiglas, optical fibers, rubber, paper—
the list reads more like an inventory of a sculptor's studio than
that of a haute couture fashion house. But Paco Rabanne
changed all that with his first manifesto-collection in February
1966. Born Francisco Rabaneda Cuervo in 1934 in the Basque
region of Spain, Rabanne fled to France with his family in
1939. From 1951 to 1963, he studied architecture at the Ecole
des Beaux-Arts in Paris, and from 1955 he helped support
his family by producing handbag and shoe designs for Roger
Model and Charles Jourdan. He continued to develop links
with couturiers by making buttons in unusual materials,
including noodles or coffee beans glued onto plastic, in the
years before unveiling his first collection, "Twelve Unwearable Dresses."

Made from plastic cut into strips and strung together with
metal rings, "Twelve Unwearable Dresses" was a sensation
and brought Rabanne international fame. In the mid-1960s,
experimentation was the order of the day, and many traditional
materials were discarded. Rabanne rejected the notion that
fabric was the only material suitable for clothes. As he wrote
in a preface to his biography (written by Lydia Kamitsis),
"far from being attached to the past, I was not only following
the spirit, but also, the art of the times. A fantastic revolution
conducted with new materials. Architecture renounced
stone, paintings canvas, sculpture marble. I, in turn, engulfed
in the flight of a changing society, abandoned my needle
and thread for my pliers and blow-torch."

It is hardly surprising, then, that aluminum was incorporated
into Rabanne's repertoire of materials. In 1966, he ordered
his first rectangles and squares of aluminum sheet. A short
time later, a company supplied Rabanne with the aluminum
disks it used to make protective aprons (worn by butchers in
abattoirs, for example). The model seated on the bench is
wearing a suit from 1967 that combines both square aluminum
sheet and prefabricated apron disks.

Paco Rabanne continues to use aluminum: his 1999/2000
couture collection included several aluminum dresses. Metal
clothing is not new—chain mail and armor have existed for
centuries—but Rabanne's reinvention and reinterpretation
using aluminum moved it from the masculine battlefield to the
feminine catwalk, and transferred the vocabulary of the tough
warrior, both protected and protective, to a new arena.

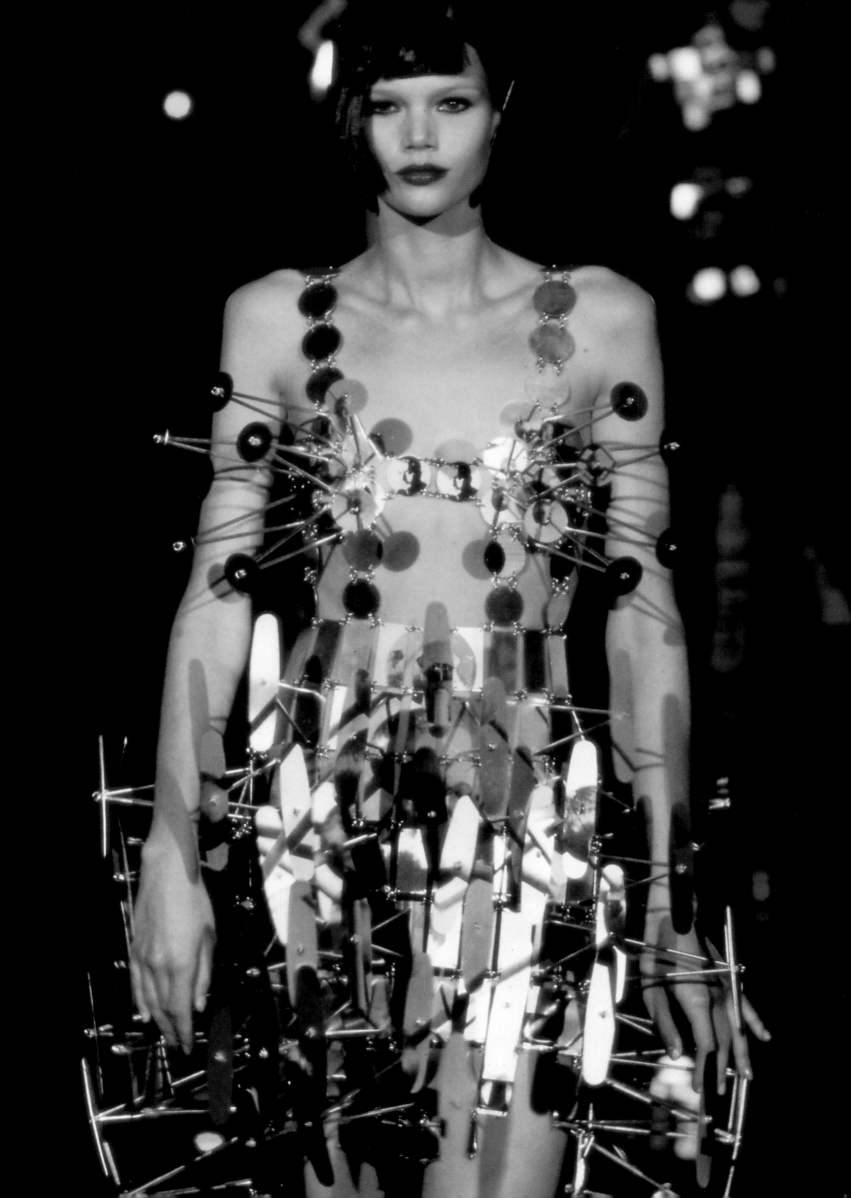

Gijs Bakker (Dutch, b. 1942)
Stove-piping necklace and armband, 1967
anodized aluminum

Arline Fisch (American, b. 1931)
Necklace, 1984
pleated and anodized aluminum
cat. 5.59

Shiang-shin Yeh
(American, b. Taiwan, 1969)
Bracelet, 1997
anodized aluminum
cat. 5.57

(opposite)
Marcia Lewis (American, b. 1946)
Creature Collar II and Stethoscope
Neckpiece, ca. 1978–80
aluminum, feathers, laminated
vegetable ivory, and ebony
cat. 1.3 and 5.55

Some of the oldest objects made of aluminum early in its history include bracelets and brooches (Highlight 2). But once the metal became cheaper, more widely available, and less a scientific marvel, it virtually disappeared from the repertoire of jewelry materials. From a design standpoint, the next significant development in aluminum jewelry occurred in Europe and America in the 1960s. No one style dominated but, rather, many individuals radically changed the way jewelry was perceived. Described as the "jewelry designer who dislikes jewelry," Dutch artist Gijs Bakker had a traditional training in goldsmithing. From the mid-1960s, he designed a collection of collars and bracelets made of aluminum. Being light-weight, malleable, and strong, the material was a practical choice, especially given the large scale of many of the collars. But its use also broke with the tradition of making important jewelry from precious materials. Bakker's work, along with that of a group of artists who followed his lead—Frans van Nieuwenborg and Martijn Wegman among them—was minimal in form and highly sculptural so as to emphasize the human body.

In the United States, artists such as Arline Fisch and Marcia Lewis began to work in aluminum in the mid-1970s. Fisch was the first American jeweler to work with aluminum and color. An advocate of applying textile techniques to jewelry, in 1982–83, Fisch began making large pleated anodized aluminum neckpieces that dissolved the distinction between ornament and dress. She continued to explore these ideas and aluminum through dramatic pleated forms and vibrant anodized colors. Lewis, a student of Fisch, credits the latter with introducing her to "bold and imaginative ways to approach metal." Like Bakker and other European jewelers, Lewis wanted to make "wearables that concerned themselves with the human anatomy and covered increasingly larger portions of it." A grant from the National Endowment for the Arts freed her to explore aluminum and the technical difficulties of working with it in a studio situation. The new skills acquired during this period of exploration enabled her to design a body of work—including the two neckpieces illustrated here—that would have been difficult to produce and impossible to wear had she used any other metal. This is true as well for Shiang-shin Yeh's precision-engineered architectonic bracelet.

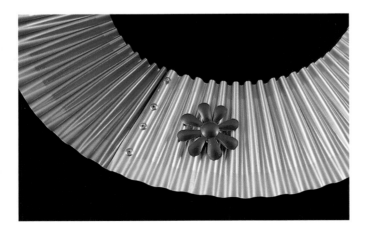

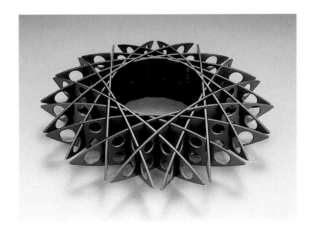

Bang & Olufsen (Danish, 1925–present)
BeoLab 4000 loudspeakers, 2000; designed
1997, manufactured 1997–present
anodized aluminum and other materials
cat. 1.8

(opposite)
**Jane Adam** (British, b. 1954)
Bangles, 1999
block-printed, dyed, and crazed
anodized aluminum

**Jane Adam** (British, b. 1954)
Earrings, 1999
dyed and crazed anodized aluminum,
gold leaf, and freshwater pearls

When exposed to air, aluminum becomes coated with a thin protective film of aluminum oxide, which gives the metal its stability and resistance to corrosion. This organic process can be enhanced by an electrolytic process called anodizing, in which the metal is coated with a hard layer of oxide by dipping it in baths of acid. Anodizing also allows for the addition of color, as dyes can be introduced into the dipping baths. The effect of a colored anodized surface is quite unlike a paint or lacquer finish, and is unique to aluminum.

Manufacturers have produced colored anodized aluminum objects since the 1920s. In 1923, Alcoa filed a patent for coloring aluminum by using dyes with anodized coatings. This method was introduced commercially in 1928 and subsequently became known worldwide under Alcoa's Alumilite trade name. Colored housewares became extremely popular in this period. "In order to determine the extent of the color idea in housewares," *House Furnishing Review* in 1927 made a "thorough canvass of the entire home equipment field" and later reported that "as bees are attracted to honey, so women are attracted to color." By September 1932, the Fanta Colored Aluminum Ware Company offered whistling kettles in eight colors, including ivory, apple green, yellow, red, and tangerine. The prewar anodized palette has been much expanded since the 1950s, as has the range of colored aluminum objects available on the market. For example, Bang & Olufsen, a Danish electronics company, opened a state-of-the-art anodizing plant in 1992 as part of a shift to anodized aluminum for their products. This eliminated the more polluting galvanizing processes needed for chromium and nickel. The company based its decision on environmental as well as aesthetic criteria.

Anodizing is also used creatively by artists. Until it is sealed (for example, with boiling water), anodized aluminum acts like porous paper. Jane Adam, a British jeweler, uses inks and felt-tipped pens to add pattern, layer upon layer, to the surface of anodized but unsealed aluminum that she purchases commercially. Anodizing requires fume extraction and precision temperature control. Such constraints limit the scale on which a studio artist can work, which is why the creative exploration of anodized aluminum has been undertaken primarily by artists working on a small scale.

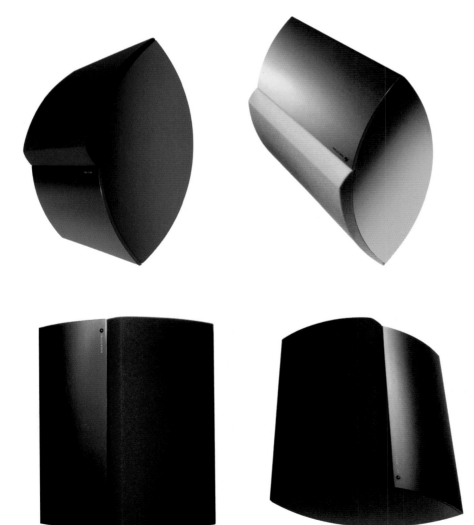

**Philippe Starck,** designer
(French, b. 1949)
Driade SpA, manufacturer
(Italian, ca. 1985–present)
Romantica chair, 1987
aluminum sheet and aluminum tubing
cat. 5.23

**Philippe Starck,** designer
(French, b. 1949)
Emeco, manufacturer
(American, 1944–present)
Hudson chair, 2000; designed
1999–2000, manufactured 2000
brushed and polished aluminum
cat. 5.22

(left) Emeco (American, 1944–present)
1006 chair, 1999; designed 1944,
manufactured 1944–present
aluminum
cat. 5.20
(right) The General Fireproofing
Company (American, 1902–present)
Chair, 1940s
aluminum
cat. 5.21

(opposite)
**Philippe Starck,** designer
(French, b. 1949)
Vitra AG, manufacturer
(Swiss, 1934–present)
W. W. Stool, 1999; designed 1990,
manufactured 1992–present
varnished sand-cast aluminum
cat. 5.25

One of the most well-known figures in contemporary design, Philippe Starck hit the headlines in 1982 when he was commissioned to design President Mitterrand's private chambers in the Palais de l'Elysée. Talent, timing, and a flamboyant media persona have kept him in the public eye ever since. His portfolio includes many designs that have become landmarks of the postindustrial age. Although Starck uses the full range of contemporary materials, aluminum figures prominently in his work, perhaps a consequence of childhood influences: his father was an aircraft designer. Another recurring motif is the streamlined, elongated horn shape, exemplified by the W. W. Stool designed in 1990 as part of a fantasy office environment for the film director Wim Wenders and put into production by Vitra in 1992. The stool is made from sand-cast aluminum, and although the pale green enamel coating negates the metallic quality of the surface, the flowing, organic form captures the material's fluidity.

One of Starck's most recent designs reflects his interest in aluminum and his willingness to work within prescribed parameters. Emeco of Hanover, Pennsylvania, began producing the all-aluminum 1006 chair for the U.S. Navy in 1944. Elements of the design hark back to Alcoa's Alcraft chair from the late 1920s. However, in the Alcoa chair the seat was not aluminum and the side rail was not welded to the back legs. The General Fireproofing Company of Ohio later produced a very similar all-aluminum chair, but again without the side rail welded to the back legs and with less sophisticated and complex curves to the stretchers.

Emeco has produced the 1006 chair continuously since 1944 using the same labor-intensive, highly skilled manufacturing techniques. In recent years, the 1006 has been taken up by design professionals and can be seen in many new restaurants and in advertising layouts. Starck used the 1006 in the newly renovated restaurant of the Paramount Hotel in New York. So when Emeco asked Starck to design a new all-aluminum chair compatible with the company's production processes and sympathetic to the original 1006, he enthusiastically accepted the project. The result: the Hudson, released in the spring of 2000.

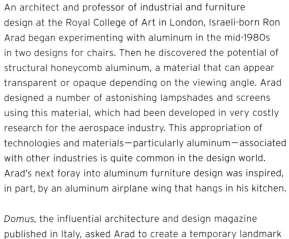

Ron Arad, designer (Israeli, b. 1951)
Ron Arad & Associates, manufacturer
(British, 1989–present)
*The "Domus" Tower*, 1997
H. 33 ft. (10 m)

Ron Arad, designer (Israeli, b. 1951)
Ron Arad & Associates, manufacturer
(British, 1989–present)
Tom Vac, 1997
aluminum and stainless steel
cat. 5.31

(opposite)
Ron Arad, designer (Israeli, b. 1951)
Ron Arad & Associates, manufacturer
(British, 1989–present)
Un-cut, 1997
anodized aluminum and
stainless steel
cat. 5.30

An architect and professor of industrial and furniture design at the Royal College of Art in London, Israeli-born Ron Arad began experimenting with aluminum in the mid-1980s in two designs for chairs. Then he discovered the potential of structural honeycomb aluminum, a material that can appear transparent or opaque depending on the viewing angle. Arad designed a number of astonishing lampshades and screens using this material, which had been developed in very costly research for the aerospace industry. This appropriation of technologies and materials—particularly aluminum—associated with other industries is quite common in the design world. Arad's next foray into aluminum furniture design was inspired, in part, by an aluminum airplane wing that hangs in his kitchen.

*Domus*, the influential architecture and design magazine published in Italy, asked Arad to create a temporary landmark in the center of Milan to promote the magazine during the 1997 Furniture Fair. The editor proposed a sculpture of one hundred stacking chairs. The resulting family of chairs tells a fascinating story about the intersections of art, design, technology, process, material, and economics.

Arad had been interested in making a chair based on concentric circles to go with a table he had designed earlier. He also wanted to use vacuum-formed aluminum. Arad's studio made a fiberglass model of the chair, which was then sent to Superform Aluminum—a company specializing in precision-engineered aluminum components that also happens to be the manufacturer of Arad's airplane wing—with two critical specifications: the chair had to be stackable and capable of being made with vacuum-formed aluminum. Superform devised an intricate steel mold for the chair. As Arad says, "Domus had a budget for sculpture, I used the budget to invest in the tool to make the chair."

The result was the Tom Vac chair. Thirty-three feet of Tom Vacs were used in *The "Domus" Tower*. A series of limited-edition and one-off chairs based on the Tom Vac were produced in 1997, including Un-cut, the untrimmed aluminum prototype version; Carbo Tom, made from carbon fiber; and Pic Chairs, twenty unique chairs made from fiberglass and resin. Tom Vac also evolved into an injection-molded plastic version for mass production made by Vitra, a long-time supporter of Arad's work, and launched at the 1998 Cologne Furniture Fair.

Junichi Arai (Japanese, b. 1932)
Blue, 1995
wool, polyester, and aluminum
cat. 5.19

Reiko Sudo, designer (Japanese, b. 1953)
Nuno Corporation, manufacturer
(Japanese, 1984–present)
Rusted Silver Washer, 1991
cotton, polyester, and aluminum lamé
cat. 5.17

(opposite)
Issey Miyake (Japanese, b. 1938)
Dress and hat from the Starburst series,
autumn/winter collection, 1998
aluminum and cotton
cat. 5.7

Contemporary Japanese designers are producing some of the most exciting and imaginative textiles and clothing in the world today, and aluminum has played a significant role in their work. Textiles embellished with metallic yarns have had a long and rich history in Japan, and those intended for kimonos emphasized the relationship between the fabric and physical movement rather than the body's form. Today, designers combine the virtues of these traditions with innovative technology. In the past, metallic yarns that could be woven into or embroidered on cloth were created by twisting metal leaf around silk or paper. Now, with synthetic fibers and new technology, metallic yarns can be produced more economically. These aluminum-coated, polyester slit films are easier to weave and also create more iridescent surface effects, as can be seen in Reiko Sudo's Rusted Silver Washer of 1991. Junichi Arai, who began weaving with metallic fibers in the mid-1950s, has developed some of the most technically interesting and experimental textiles of the 1980s and '90s. He has obtained patents for the process of vacuum-plating polyester film with aluminum, and he also invented the method of dissolving some of the metal in an alkali solution once the fabric is woven. This technique, known as "melt-off," leaves behind a semitransparent cloth with contrasting surface textures.

Historically, another method of producing metallic, reflective fabrics was to apply metallic leaf directly to the cloth to fill in areas between heavily embroidered motifs. Issey Miyake has incorporated this method into his vision for 21st-century clothing. His Starburst designs, which caused a sensation when shown in his 1998 autumn/winter collection in Paris, express in a concrete and proactive way his feelings about environmental issues. Miyake believes that "it is crucial to revise manufacturing and production processes, and to stimulate a new inventiveness in textile techniques by recycling fabrics, whilst still creating magical clothes with free, timeless forms." In the Starburst series, clothes are heat-pressed between silver-, bronze-, or gold-colored aluminum foil. When worn, the pressure of the body causes the foil to crack and stretch, and the garment literally bursts through the aluminum. Miyake has presented us with a spectacular way to reuse or breathe new life into clothing. Starburst is an idea-in-process; in many ways the task of completion is in the public's hands.

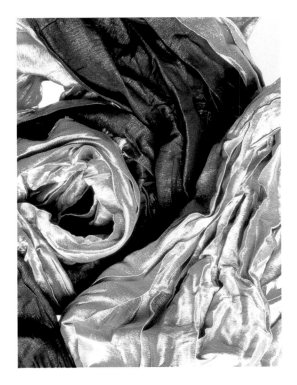

**Marc Newson,** designer
(Australian, b. 1963)
Biomega, manufacturer
(Danish, 1998–present)
MN-01, Extravaganza bicycle, 2000;
designed 1998, manufactured
1999–present
superplastic aluminum and
other materials
cat. 4.44

**Marc Newson,** designer
(Australian, b. 1963)
Pod, manufacturer
Orgone chair, 2000; designed 1993
polished aluminum and paint
cat. 5.29

(opposite)
**Marc Newson,** designer
(Australian, b. 1963)
Pod, manufacturer
MN-01 LC1, Lockheed Lounge,
1986–88, designed 1985
riveted sheet aluminum over
fiberglass and rubber
cat. 5.64

Since his first exhibition in 1985, which included the now iconic Lockheed Lounge, Marc Newson's designs have been media events. Newson personifies the new generation of multidimensional, "spoon to city" designers who are rapidly becoming household names. He has designed everything from cars, bikes, restaurant interiors, and furniture, to toilet-roll holders, bottle openers, and dish drainers, making use of a variety of materials, including aluminum. He studied jewelry and silversmithing at art school because it was "the one department which taught you how to build stuff," but it also ignited his love of technique and rigorous attention to detail.

The sensual, curvaceous shape of the Lockheed Lounge is based on Newson's vision of a "fluid metallic form, like a giant blob of mercury," which has become the keystone of his design vocabulary. Named after the American aircraft manufacturer, the chair's body is sculpted from fiberglass-reinforced plastic. The surface is covered with thin, nonoverlapping aluminum sheets joined with blind rivets, giving the impression of an airplane fuselage. Form and materials overwhelm any sense of comfort and coziness. The Lounge, produced in a limited edition of ten, secured a place in popular and design culture when Madonna used it in her 1988 *Rain* video and Philippe Starck integrated it into the 1990 redesign of the Paramount Hotel foyer in New York.

The double blob reappears in Newson's Orgone furniture series, which again used aluminum. Here, the aesthetics and technology associated with the material dominate. In the Orgone chair, intense handworking illustrates the transference of skills and technologies that occurs frequently in contemporary design. The chair is made using traditional car-panel beating techniques—rolling, hammering, and welding of thin sheet aluminum—by master craftsmen who once produced the coach work for Aston Martin and now specialize in repairing the bodywork of luxury sports cars. The Orgone chair is also available in plastic in a wide variety of colors at a considerably lower price.

A recent Newson design is the aluminum-framed urban bicycle for Biomega, an environmentally conscious Danish company that wants to redefine biking and, in the process, get city dwellers out of cars and onto bicycles. The company recognizes that, in part, people acquire cars to make a statement, but believes that a bike designed by Newson can play the same role.

African
Fon staff, possibly 1930s
aluminum, iron, and enamel inlay
cat. 4.62

African
Beads, 20th century
aluminum
cat. 1.2

Gabbra women wearing aluminum
necklaces, Kenya, ca. 1980

(opposite)
African
Asipim chair, 20th century
wood, aluminum, and steel
cat. 4.61

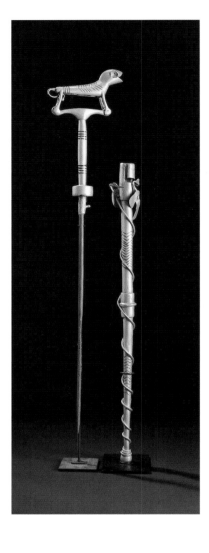

Aluminum is malleable and can be worked relatively easily even when cold. It also melts at a much lower temperature (660°C) than metals such as silver (961°C) and iron (1535°C), which makes it an ideal candidate for recycling on an industrial scale. But in Africa, aluminum has been recycled and reworked at the level of the individual village smith or ornament maker for decades. African craftsmen initially obtained aluminum from cooking pots and other goods introduced by missionaries and colonists. In the 1930s, France, for example, actively promoted aluminum furniture, household goods, and architecture in its African colonies through specialist exhibitions and fairs held in Paris. An object that may date from this period is the aluminum sword-cum-staff made by a Fon artist from the Republic of Benin, where elaborately decorated weapons served ceremonial rather than functional purposes.

In Kenya, aluminum is the most popular metal for ornaments and jewelry, including beautifully faceted beads like the dramatic multistrands worn by the Gabbra women shown here. The appeal of the metal lies in the fact that it looks like silver but is much cheaper and does not tarnish. In the 1970s, anthropologist Jean Brown spent many months documenting traditional Kenyan metalworking practices. Smithing is a highly specialized, skilled, and protected trade. Ornament makers, on the other hand, are not allowed to maintain permanent workshops, nor can they use the traditional tools of the smith, including bellows for the fire. Even charcoal is off-limits. But because aluminum is so easy to rework, ornament makers can fabricate objects with it even under such constraints.

In Ghana, the Asante have produced three types of chair designs based on European forms. The one illustrated here is the *asipim*, usually decorated with nails, knobs, and finials, and occasionally with sheet metal. In this instance, the wooden frame is covered with aluminum and embellished with steel nails or studs. In Asante culture, chairs are believed to house the owner's soul, thereby acquiring a spiritual significance in direct proportion to the social status of the owner. Thus the humble aluminum cooking pot that is recycled into a chair is utterly transformed in its new society and culture.

Clare Graham (American, b. 1949)
*Carpet of Printed Can Labels*
and *Serpentine Chaise Longue*,
detail and full view, 1997
reused aluminum beverage cans
cat. 4.66 and 4.67

(opposite)
Boris Bally (American, b. 1961)
Transit chairs, 1997
reused aluminum traffic signs
cat. 4.64

The secondary aluminum industry—which obtains the metal by melting scrap or recycled aluminum—first developed about 1912 and was well established soon after the start of World War I. By 1937, according to *Aluminium and Electrolysis*, it was "one of the foremost branches of the general aluminium industry." Yet its scale and economic, social, and environmental impact were minor compared to today's industry. For example, U.S. figures for total recovery of aluminum in thousands of metric tons were 46.6 in 1935 and 3,685 in 1997. Recycling aluminum is economically viable for three reasons. First, recycling aluminum requires only 5 percent of the energy necessary for primary production. Second, aluminum can be resmelted indefinitely without deterioration of its intrinsic properties. And third, effective collection schemes exist today because recycling is ecologically expedient, politically encouraged, and embedded in the consumer's psyche.

Designers, architects, and craftspeople incorporate used aluminum products in their works for a variety of reasons— aesthetic, economic, social, and political. Clare Graham is an avid collector of things that often end up in the garbage, from swizzle sticks and bottle caps to the pull tabs of beverage cans and the cans themselves. Graham admits to an obsession with quantity and multiples. His private vision transforms his vast collection of objects with little intrinsic value and questionable aesthetic appeal into unique, fantastic forms. Boris Bally, a skilled metalsmith, began using recycled aluminum road signs in 1991. Initially he produced bowls, and because the aluminum could not be heated without destroying the enameled surface, he raised it by hand hammering. He later employed more commercial spinning and machine-bending techniques, still with reused signs as the basic material, first in the production of bowls and trays, then in a chair design known as the Transit chair, which was patented in 1997.

Bally forages up and down the eastern United States in search of street signs, rescuing them from the recycling mill by offering scrap dealers more than the going rate for recycling. One recent haul brought in 10,000 pounds of signs from Ohio. The work of Graham, Bally, and other artists who use waste material will not correct the imbalance between production and consumption of natural resources. It does, however, raise awareness about recycling and provide a new future for objects that might otherwise have had an ignominious end in the smelter.

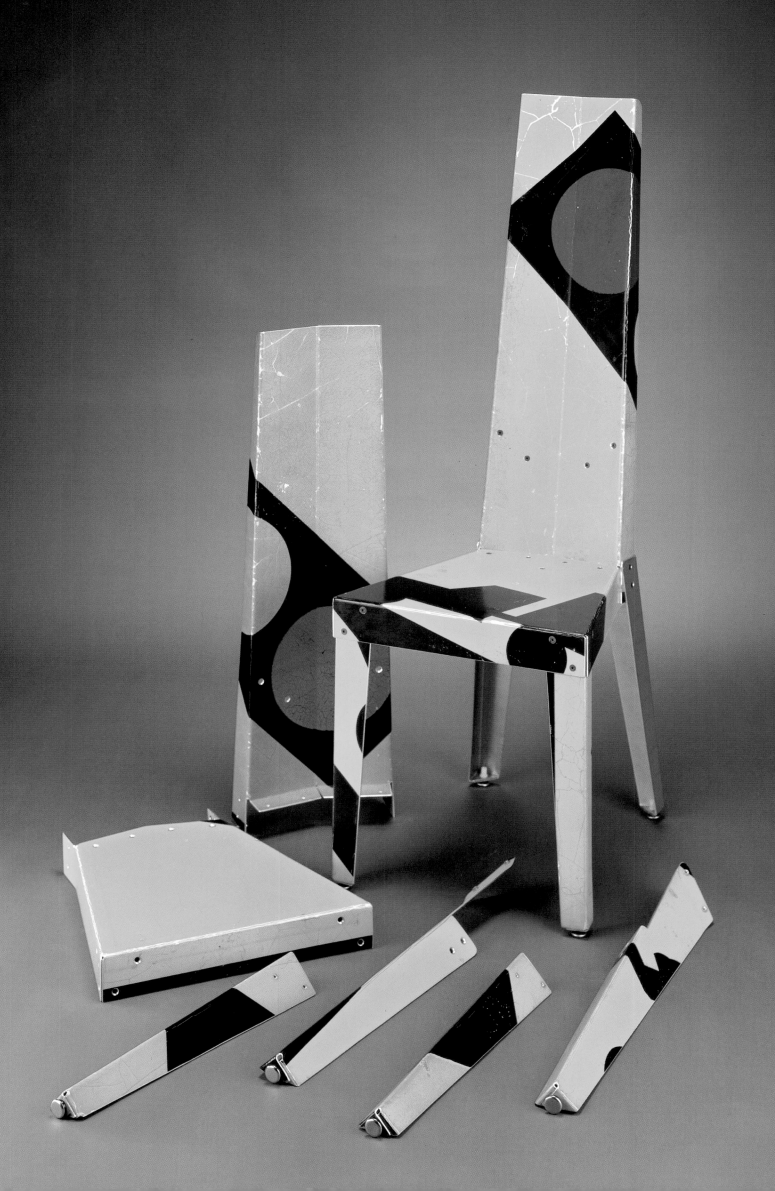

Bořek Šípek, designer (Czech, b. 1949)
Vitra AG, manufacturer
(Swiss, 1934–present)
Sedlak chairs, 1992
aluminum, beech, and molded
polyurethane
cat. 4.71

Philippe Starck, designer
(French, b. 1949)
Vitra AG, manufacturer
(Swiss, 1934–present)
Stacked Louis 20 armchairs, designed
1991, manufactured 1992–present
polished aluminum and blown
polypropylene
cat. 4.72

(opposite)
J Mays (American, b. 1954)
et al., designers
Drawing of multisport pickup
concept car, 1995–97
ink on paper

J Mays (American, b. 1954)
et al., designers
Drawing of multisport utility
concept car, 1995–97
ink on paper

Shipping furniture as parts "knocked down" to be assembled at their destination has been a common cost-saving practice among furniture makers for centuries. In recent years, however, with the emphasis on environmental protection and recycling, the practice has grown into a "design for disassembly" approach that takes into account an object's end as well as its beginning. Design for disassembly is becoming standardized as manufacturers across many industries take more responsibility for the disposal of their products, and this approach is especially suitable for objects that contain aluminum, because the metal can be easily and profitably recycled (Highlight 39).

One strategy is to "build in" an object's recyclability by making its components easily separable. Philippe Starck's Louis 20 armchair consists of aluminum arms and back legs that are screwed to the polypropylene seat and back. It can be dismantled into recyclable elements in a matter of seconds simply by undoing the screws. The same is true of Bořek Šípek's Sedlak chair, with its aluminum back stiles. The concept of design for disassembly has also generated some creative thinking about how to extend an object's life. For example, between 1994 and 1997, Alcoa and a design team led by J Mays, now head of design at Ford, developed three multisport concept vehicles—a pickup, sedan, and utility vehicle. Three innovative ideas guided the team's work: technical modularity and styling flexibility—both made possible by the use of the aluminum space frame—and leveraging the "aluminumness" of the vehicle, that is, making visible its aluminum content. All three vehicles use the same aluminum space frame apart from the rear nodule. Changing this nodule creates a different profile and a different vehicle, which means that the manufacturer can produce three different cars on one assembly line rather than three. The owner can also alter the vehicle to accommodate family and lifestyle changes. Exterior aluminum panels fit to the frame and can be easily replaced, either to make repairs or to change the appearance of the car. The front end can also be modified: three versions of an interchangeable aluminum "clip" allow the owner to create a new look for the vehicle at will. The extensive use of aluminum in these models has created new possibilities in assembly and disassembly, not only at the beginning and end of a car's existence but also during its useful lifetime.

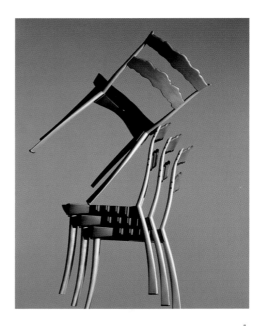

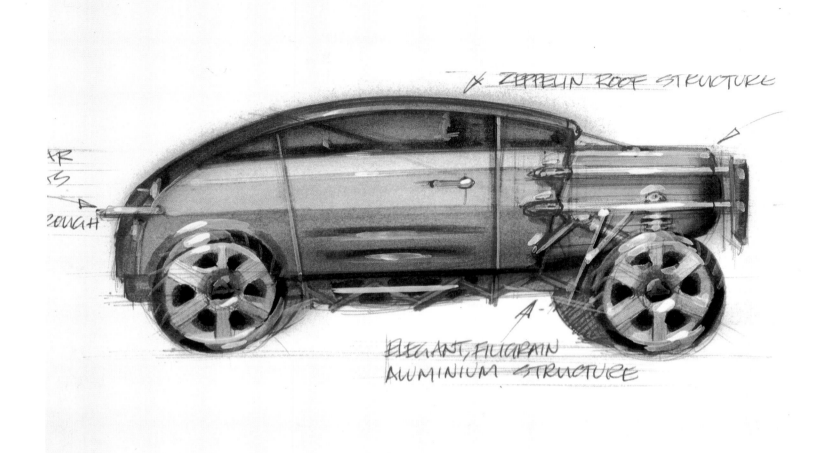

*⊀ ZEPPELIN ROOF STRUCTURE*

*⊀R*
*S*
*⊀OUGH*

*ELEGANT, FILIGRAIN*
*ALUMINIUM STRUCTURE*

*...RS DISAPPEAR*
*...E A-PILLARS*
*...IN THE*

Carnegie Mellon University
Field Robotics Center
(American, 1979–present)
Dante II, 1994
aluminum and other materials

Carnegie Mellon University
Field Robotics Center
(American, 1979–present)
Nomad, 1997
aluminum and other materials

(opposite)
Sony Electronics Inc.
(Japanese, 1946–present)
Aibo robotic dogs, 1999
aluminum, plastic, and other materials
cat. 5.61

Carnegie Mellon University
Robot Learning Laboratory
(American, 1990–present)
Jeeves, the Tennis Rover, with and
without cover, 1996
aluminum, plastic, and other materials
cat. 5.62

Aluminum is used in robotics today for the same reasons it was employed in precision instruments at the end of the 19th century (Highlight 5). Often the aluminum is not visible but comprises the robot's interior structure: both Jeeves and Aibo, for example, have plastic exteriors but aluminum parts inside. Advances in robotics have impacted society in numerous ways, from industrial robots in manufacturing plants to planetary explorers such as Sojourner, part of the Mars Pathfinder mission in 1997. Dante II was designed to negotiate the sheer crater walls of volcanos, collecting information on volcanic gases, as well as to demonstrate new robotic techniques for exploring extreme terrains such as those found on extraterrestrial surfaces. Recently, robots have entered the domestic environment, not as the household help envisioned by science fiction, but as entertainment items like Sony's robotic dog, Aibo. However, the technology behind Aibo—the name derives from "artificial intelligence" and "robot"—has great potential for other applications. Aibo is an autonomous robot that makes its own decisions and acts according to its own programmed capacity for emotion, instinct, learning, and growth, all key factors in the latest developments in "intelligent" robots. Jeeves, a tennis-ball-collecting, "thinking" robot, was designed, built, and programmed over a six-month period by the Robot Learning Laboratory at Carnegie Mellon University in Pittsburgh, and won first prize in an artificial intelligence robot competition in the category "clean up the tennis court."

Nomad, another intelligent machine made of aluminum and other materials, is programmed to search rock-strewn areas of Antarctica for meteorites, which requires the ability to distinguish between terrestrial material and objects that fall from space. Nomad arrived in Antarctica by helicopter: its light-weight aluminum components and resistance to corrosion and harsh environments helped make it fit for its duties. NASA, which sponsored the research into Nomad by the Field Robotics Center at Carnegie Mellon University, hopes eventually to send intelligent robots to other planets. Because two-way radio communication between space missions and scientists on Earth can take hours, it would be impossible to guide a robotic explorer. For this work, robots will need the ability to make decisions for themselves.

Chrysler Corporation,
now DaimlerChrysler Corporation
(American, 1924–present)
Prowler, manufactured 1996–present
aluminum and other materials
cat. 5.65

Jaguar Cars (British, 1921–present)
Jaguar XJ220, manufactured 1992–94
aluminum and other materials

(opposite)
Audi (German, 1909–present)
Audi A8, manufactured 1994–present
aluminum and other materials

Audi (German, 1909–present)
Audi A8 4.2 quattro space frame,
manufactured 1994–present
aluminum
cat. 4.51

In 1960, the average amount of aluminum used in a car was 63 pounds. By 1999, that average had reached 241 pounds. In recent years, aluminum producers have worked closely with car manufacturers (for example, Alcoa with Audi, Daimler-Chrysler, and Ferrari, and Alcan with Jaguar, now owned by Ford) to revolutionize the way aluminum is used in vehicles and, in the process, to rethink car design. These high-profile collaborations represent a strategy for promoting automotive applications of aluminum to other car manufacturers as well as the public.

Changes were first seen in the extreme luxury and specialist end of the market with the Jaguar XJ220, produced in 1992–94 with an original price tag of $706,000. The car's structure—an aluminum-honeycomb chassis bonded with adhesives, an aluminum body, and numerous other parts—resulted in superior rigidity and a dramatic decrease in weight. The design mandate behind the Ferrari 360 Modena was simple: reduced weight, increased performance, better handling, stiffer structure, more interior room and comfort—and a minimal increase in price. The key to achieving all of this was aluminum, which accounts for about 1,240 pounds of the car's total weight of 3,100 pounds. Another aluminum-intensive car, the Plymouth Prowler—"a salute to hot rodding"—developed out of a 1993 partnership between Alcoa and Chrysler (now DaimlerChrysler) formed to investigate new materials and manufacturing processes.

The Audi A8, produced in 1994 and introduced to the United States in 1997, utilized the ASF (Audi Space Frame) concept developed by Audi and Alcoa over a ten-year period; in the process, seven new aircraft-grade aluminum alloys were created. In space-frame construction, aluminum body panels are integrated into a high-strength aluminum frame structure that consists of extruded sections connected by vacuum die-cast nodes. The aluminum body is 40 percent lighter than the steel body of a comparable car, which results in greater fuel efficiency and performance figures. The ASF has since been refined to reduce the number of individual components, thus saving on costs for labor and parts and further decreasing weight. When introduced, the Audi A8 was described as "the first luxury car of the 21st century." At the 1999 Frankfurt Motor Show, Audi unveiled the A2, the first high-volume passenger car with an all-aluminum body. Aluminum car technology developed at the end of the 20th century is shaping the way many of us will drive in the 21st.

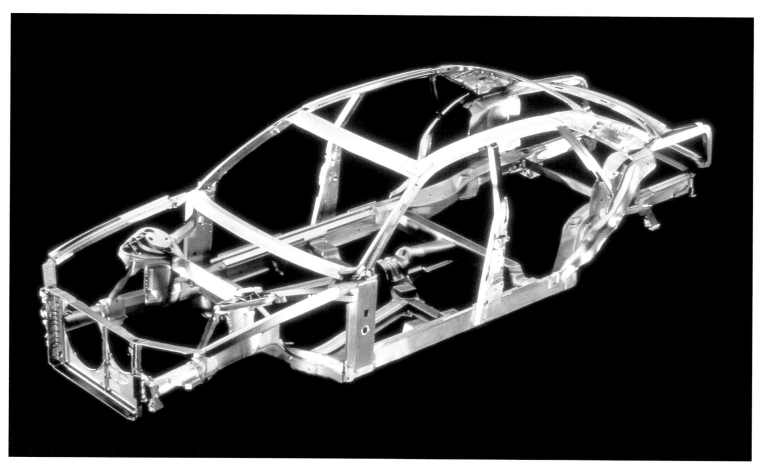

This glossary contains technical definitions for aluminum, its fundamental composition, and the methods by which it is commonly processed. A variety of sources were consulted in order to present more interesting and informative definitions than a list of technical terms might offer. In some cases, definitions contemporary with the invention of a process or product have been included. Multiple definitions have been cited to clarify certain terms. Whether read discretely or as a whole, this glossary provides another chapter in the presentation of aluminum's history.

## Key to Sources

**ACT** = *Aluminum Casting Technology*, ed. D. Z. Zalensas (Des Plains, Ill.: American Foundrymen's Association, 1993)

**AL** = INFALUM (Informationsstelle für Aluminium und Umwelt), *Aluminum Lexicon* (Oberbözberg, Switzerland: Olynthus Verlag, 1990)

**ANL** = *Aluminum News-Letter* (Pittsburgh: Alcoa, 1934–46)

**Bronzes** = "Aluminum Bronzes: Physical and Chemical Properties," *Aluminum World* 6, no. 6 (March 1900): 111

**DA** = Mark Firth and Louis Skoler, *The Dictionary of Art* (New York: Grove Dictionaries, 1996)

**Dirt** = *It all starts with dirt: the making of aluminum at Alcoa* (poster) (Pittsburgh: Alcoa, 1998)

**Dobell** = Mercy Dobell, "Earn More by Learning More about Multi-Use Conveniences and Advantages of Cooking and Serving Utensils," *House Furnishing Review* 109, no. 3 (September 1948): 51–59

**EA** = INTEXALU Group, *Encyclopedia of Aluminium* (www.intexalu.com)

**EncyA** = *Encyclopedia Americana*, international ed. (Danbury, Conn.: Grolier, 1992)

**EPST** = *Encyclopedia of Physical Science and Technology* (Orlando, Fla.: Academic Press, 1987)

**FPI** = Formed Plastics Industries, Alpharetta, Georgia (www.fpind.com)

**NCE** = *New Columbia Encyclopedia*, 4th ed. (New York: Columbia University Press, 1975)

**New World** = *Webster's New World Dictionary of the American Language*, 2d college ed. (New York: Simon & Schuster, 1980)

**RR** = Rolf Rolles, Light Metals Research Bureau, New Kensington, Pennsylvania

**TALAT** = Geoff Budd, "Resources and Production of Aluminum," TALAT Lecture 1100 (Birmingham, Ala.: Aluminum Federation, 1994)

**Webster's Third** = *Webster's Third New International Dictionary of the English Language* (Springfield, Mass.: Merriam-Webster, 1986)

**alloy.** A substance composed of two or more metals intimately mixed and united usually by being fused together and dissolving in each other when molten. (Webster's Third)

A substance with metallic properties, composed of two or more chemical elements of which at least one is a metal. More specifically, aluminum plus one or more other elements produced to have certain specific, desirable characteristics. (Dirt)

**alum/alumen.** Either of two colorless or white isomorphic crystalline double sulfates of aluminum having a sweetish-sour astringent taste and used chiefly in medicine, internally as emetics and locally as astringents and styptics: a) the potassium double sulfate $KAl(SO_4)_2 \cdot 12H_2O$ occurring naturally and also made commercially (as by treating bauxite with sulfuric acid and then potassium sulfate): potassium aluminum sulfate—called also potash alum, potassium alum; b) the ammonium double sulfate $NH_4Al(SO_4)_2 \cdot 12H_2O$ made commercially (as from ammonia sulfate and aluminum sulfate)—called also ammonia alum, ammonium alum. (Webster's Third)

*Alumen* is the Latin name for alum from which the name aluminium is derived. This name was adopted in the early 19th century and is used worldwide except in the United States, where the spelling is *aluminum*, and in Italy, where *alluminio* is used. (DA)

**alumina.** Aluminum oxide or alumina is an extract of bauxite and is the raw material of electrolyzed aluminum. Its appearance is of a fine white powder; alumina is a hard compound with only diamonds and a few other synthetic products having superior strength. At room temperature, alumina is insoluble in all known chemical compounds and is a poor conductor of electricity. It has a high fusion temperature (2,000°C). According to Karl Joseph Bayer's method (1887), bauxite ore is pulverized, mixed with soda, and heated under great pressure at a high temperature. The obtained liquid solution, sodium aluminate, is purified, diluted, and cooled, which provokes the precipitation of hydrated aluminum oxide. This then undergoes calcination to obtain alumina, fundamental in the production of aluminum. Worldwide production of alumina surpasses 43 million tons per year. The most important alumina plant is located in Gladstone, Australia. Its production capacity is 3.3 million tons of alumina per year. In addition to the production of aluminum, alumina is also used in the manufacturing of pottery, glass, chinaware, and spark plugs, and plays an important role in chemical water treatments. (EA)

**aluminium.** See aluminum

**aluminum.** A bluish silver-white trivalent metallic element, very malleable, ductile, and sonorous and noted for its lightness, good electrical and thermal conductivity, high reflectivity, and resistance to oxidation, that is the most abundant metal in the earth's crust, of which it forms over 7 percent, always occurring in combination (as in bauxite, cryolite, corundum, alumite, diaspore, turquoise, spinel, kaolin, feldspar, mica), that is manufactured by electrolysis of a solution of alumina in molten fluorides, followed sometimes by electrolytic refining, that is used usually in the form of alloys for structural purposes (as in construction of aircraft, automobiles, and buildings), in the chemical and food processing industries, in cooking utensils, and in electrical conductors, and that is used in the form of powder or flakes in pigments, pyrotechnic compositions, and explosives—symbol Al. (Webster's Third)

Aluminum is obtained by the electrolysis of alumina, a method discovered by the Frenchman Paul Héroult and the American Charles Hall. The method used to obtain raw aluminum consists in reducing by electrolysis dissolved alumina in molten cryolite, at a temperature of 1000°C. The vats lined in carbon (the cathode) are passed through with a high intensity electrical current. The aluminum settles on the bottom of the vat, whereas the oxygen reacts with the carbon anodes and emits the $CO_2$ gas. Automatically and at regular intervals a hole is pierced with a perforator into the solidified crust which forms at the surface of the vat. A measured quantity of alumina is then introduced into the electrolyzed vat. During the electrolyzing of the alumina, a liquid state of aluminum settles onto the bottom of the vat. It is siphoned off at regular intervals and transported by trolley to the foundry where it is emptied into an oven. Various metals of precise proportions are added to obtain the alloys with the specific qualities required. The aluminum is then regularly de-gassed before it solidifies, ready to be used in its various shapes and forms. (EA)

As the most abundant metallic element it is estimated to form about 8 percent of the solid portion of the earth's crust. It is an important constituent of practically all common rocks except sandstone and limestone; even in these it is usually present as an impurity. Aluminum has a strong affinity for oxygen and never is found in the purely metallic condition. It occurs chiefly in chemical combination with oxygen or silicon, although there are large deposits of aluminum chemically combined with phosphorus and sulfur. The commercially important ore of aluminum is bauxite. (EncyA)

One of the chemical elements, a silvery, lightweight, easily worked metal that resists corrosion and is found abundantly, but only in combination: symbol, Al; at. wt., 26.9815; at. no., 13; sp. gr., 2.699; melt. pt., 660.1°C. (New World)

**aluminum bronze.** A pale gold–colored alloy composed of copper and usually 5 to 10 percent aluminum with iron, nickel, and tin usually being present in amounts of less than one percent each and used especially for corrosion-resistant parts, for wear-resistant bearings, bushings, gears, and dies, and for ornamental articles. (Webster's Third)

The name aluminum bronze includes a variety of aluminum-copper alloys in which the copper predominates. The name aluminum bronze does not denote, as is often believed, a certain alloy, but includes a large number of alloys that vary in their properties from the greatest ductility to the greatest hardness. In using aluminum bronze it is therefore important to select the right alloy for the desired purpose. (Bronzes)

**aluminum foil.** Aluminum ingot rolled to less than .005 inch. Thickness of the foil depends on its industrial requirement. (ANL, June 1935: 5)

**aluminum sheets.** Panels or sheets of aluminum are obtained by lamination and are used in the manufacture of products with high-quality finishes, lightweight properties, and that are mechanically functional. Obtained by first fusion (or from recycling), aluminum sheet is introduced to high-temperature lamination followed by low-temperature lamination. This enables the product to achieve a thickness of as little as 6 microns (aluminum foil). Aluminum sheets of various thicknesses are used for aircraft construction, tanks, and packaging. (EA)

**anodize/anodizing.** An electrochemical oxidation process usually carried out in an acid bath. In a direct current circuit, the aluminum serves as the anode. Commercially pure aluminum produces clear oxide coatings. When alloying elements, like silicon, copper, or manganese, are incorporated into aluminum, the color of the oxide coating will be gray, bronze, or black. Anodic oxide films are porous and can also be colored with organic dyes. Exposure of the anodized coating to steam or boiling water seals these pores. Anodic coatings are used as decorative coatings on jewelry or cookware or as decorative/protective coatings on building facades or curtain walls. (RR)

A surface treatment specifically used for aluminum and produced by electrolysis. It gives a glossy metallic or satiny finish, and encompasses a wide selection of colors. It also produces a hard film resistant to exterior elements. (EA)

**antifriction metal.** Alloy used in plain bearings. Antifriction metals such as Babbitt metal and white metal are made of tin, lead, antimony, zinc, and copper in various combinations and proportions. (NCE)

**bauxite.** Bauxite ore is the mineral most widely used to obtain alumina, the intermediary material, fundamental in the manufacture of aluminum. Bauxite was first discovered by Pierre Berthier in 1821, and was named after the town in which it was found (Baux-en-Provence). This important ore of aluminum is composed of hydrated aluminum oxide (40 to 60 percent) amalgamated with silica and iron oxide, which gives bauxite its characteristic red color. Bauxite is formed by the rapid weathering of granite rocks under hot, humid climatic conditions. It is for this reason that bauxite is found in large quantities in Australia, Guinea, and Jamaica. Bauxite deposits amount to 20 billion tons and are found throughout the world. Bauxite production increases each year, currently totaling 115 million tons. Since 1991, due to the exhaustion of bauxite mines, France has practically ceased the extraction of bauxite. (EA)

Bauxite may be brown, yellow, pink, red, or a mixture of colors. It consists of chemical combinations of aluminum, oxygen, and water, called hydrated oxides of aluminum, along with oxides of iron, silicon, and titanium, which are considered impurities. Bauxite contains about 50 to 60 percent aluminum ore (aluminum oxide, $Al_2O_3$). It takes about two pounds of bauxite to make a pound of alumina, and two pounds of alumina to make a pound of aluminum. The principal sources of bauxite in the United States are in Arkansas and other southern states. Large deposits also are found in Jamaica, Russia, Greece, Guinea, France, Surinam, Guyana, Brazil, Hungary, Yugoslavia, Indonesia, and other parts of the world. (EncyA)

**brazing.** Joining metals by flowing a thin layer of molten, nonferrous filler metal into the space between them. (Dirt)

**casting.** The process of forming molten metal into a particular shape by pouring it into a mold and letting it harden. (Dirt)

It is the oldest and simplest (in theory if not in practice!) means of manufacturing shaped components. The metal is melted and poured into molds of the required shape. Molding in sand is one of the oldest industrial arts and still practiced extensively. Other processes include gravity (or permanent mold) die-casting, in which the aluminum flows by gravity into a metal mold, and pressure die-casting, in which molten metal is forced under pressure into a steel die. (TALAT)

**casting alloys.** Specific to aluminum, they generally contain larger percentages of alloying elements than wrought alloys. The principal elements added to aluminum for casting alloys are copper, silicon, magnesium, zinc, and iron. For most large castings—such as crankcases for internal combustion engines, architectural forms, or machine elements—the sand-casting process generally is used. Permanent mold castings are made by pouring the molten metals into molds. The rapid chilling provided by contact with the metal mold improves the mechanical properties of the casting. (EncyA, s.v. "castings")

**clip jointing.** A common assembly technique used in aluminum joinery for window fixtures, joints, and protective covering. This technique enables easy handling in windowpane replacements and for assembling and dismantling in aluminum construction and joinery. (EA)

**cold mill.** The equipment on which aluminum is rolled into sheet or foil by passing it through pairs of rollers under pressure. In cold rolling, the incoming metal is normally at room temperature. (Dirt)

**corundum.** Aluminum oxide $Al_2O_3$ occurring in nature in massive form and as variously colored rhombohydral crystals including the gems ruby, sapphire, oriental amethyst, oriental emerald, and oriental topaz, synthesized both in gem and industrial quality, extremely tough and with a hardness exceeded only by a few substances (as silicon carbide and diamond), and used industrially as an abrasive (hardness 9, sp. gr. 3.95–4.10). (Webster's Third)

**cryolite.** A mineral consisting of sodium-aluminum fluoride $Na_3AlF_6$ found in Greenland; usually in white cleavable masses of waxy luster and used in making soda and aluminum (hardness 2.5, sp. gr. 2.95–3.0). (Webster's Third)

**die-cast aluminum.** Die castings are made in metal dies by forcing the molten aluminum into the dies by means of a piston or compressed air. Die castings can be made with thinner sections than are practical by the sand and permanent mold processes and are so accurately sized that most machining operations are eliminated. Precision castings with excellent surfaces and high dimensional accuracy can be made in plaster molds. Aluminum castings have been produced with a weight as great as 7,000 pounds (3,175 kg) and as little as 1 ounce (28 g). (EncyA, s.v. "castings")

It is made by forcing molten metal into a mold under pressure. This method is frequently used because it is fast and reasonably inexpensive. It is also used when extreme dimensional accuracy is required. For example, waffle grids are die cast. (Dobell)

**drawn aluminum.** Drawing of aluminum enables the production of various profile sections and shapes adaptable to the many uses of aluminum such as in the manufacturing of chassis frames, aircraft wings, structural bodywork of vehicles, construction, oxygen bottles, and aircraft landing gear. The die-caster is a hollowed tool machine that gives form and shape to the profile sections. Its handling requires exacting and precise work. The tool is regulated to 100th of a millimeter but the ultimate adjustment is carried out manually. Another drawing process consists of extruding billets of aluminum by means of a hydraulic press. Cylindrical billets are passed through the die. The lengths of the billets are approximately 50 meters and provide for one or more profiles. The billets are then cut to the required dimensions for individual customers and uses. There are also profiles made of a wide range of shapes to make for easy adaptation to modern techniques. Aluminum profiles are most widely used in many sectors of industry, for example, in building, transport, and aeronautics. (EA)

**drawn ware aluminum.** This is also called "pressed aluminum" and is made from heavy sheet aluminum. During the process, the utensil is put through a series of stamping presses. Each machine draws or presses the utensil nearer to its desired shape. Pans of this type vary in quality from very good to very poor. Utensils made from "heavy gauge" drawn aluminum may be almost as thick and heavy as cast aluminum. However, cast aluminum is graduated, whereas drawn ware tends to be uniform in thickness. Heavy drawn ware with close-fitting covers may be used for waterless cooking. Drawn aluminum is especially adapted for oven cooking. Baking and roasting pans are made of aluminum. (Dobell)

**duralumin.** Alloy of aluminum (over 90%) with copper (about 4%), magnesium (0.5%–1%), and manganese (less than 1%). Before final heat treatment the alloy is ductible and malleable; after heat treatment a reaction between the aluminum and magnesium produces increased hardness and tensile strength. Because of its lightness and other desirable physical properties, duralumin is widely used in the aircraft industry. (NCE)

**electrolysis.** The process of producing chemical changes by passage of an electric current through an electrolyte (as in a cell), the ions present carrying the current by emigrating to the electrodes where they may form new substances (as in deposition of metals or liberation of gases). (Webster's Third)

**engineered product.** A basic aluminum fabricated product that has been mechanically altered to create special properties for specific purposes; forgings and extrusions are examples of engineered products. (Dirt)

**extrusion.** The extrusion process consists of forcing metal, which is heated to increase its plasticity, through a die to obtain many sections that because of their intricacy cannot be rolled and hence are not available in metals which cannot be extruded. The outstanding advantage of the extrusion process therefore is that it permits the production of shapes in which a metal is most efficiently distributed from the standpoint of engineering design. The standard rolled structural shapes are a compromise between the dictates of design and the limitations of the rolling process. The extrusion process, on the other hand, gives the designer almost a free hand to dispose the metal about the central axis to withstand the applied stresses with a minimum weight of material. Thus in addition to the saving in weight that results from their low specific gravity, aluminum extruded

sections offer greater efficiency. Another advantage of aluminum extruded shapes is in the ease of making assemblies. Because they can be tailor-made to the designers' needs, the sections can be produced to fit together with others in the most efficient way. Finally, because interlocking sections can be produced, ordinary means of joining such as welding and riveting may in most cases be eliminated. This feature reduces much of the fabrication cost of the finished structure. Most of the strong alloys of aluminum can be extruded. (ANL, June 1936: 4)

**fabricate/fabrication.** To work a material into a finished state by machining, forming, or joining. (Dirt)

**flat-rolled products.** Aluminum plate, sheet, or foil products made by passing ingot through pairs of rolls. By moving the rolls closer together and passing the ingot between them, the thickness is reduced and the length is increased. (Dirt)

**forging.** A metal part worked to predetermined shape by one or more processes such as hammering, pressing, or rolling. (Dirt)

**Hitchner vacuum sand-casting process.** A thin-wall mold is prepared from chemically bonded sand. The resin-bonded mold is porous and allows air to pass through it. The mold is placed in an airtight container and a vacuum is pulled on the container, evacuating the sand mold. Liquid metal is poured into the mold; the metal is sucked against the thin sand-wall mold and freezes quickly, forming a shell. (RR)

**honeycomb aluminum.** Honeycomb structures generally are comprised of two thin sheets of material separated by a core of foil that has been formed and is adhesively bonded to the face sheets. (The thin direction of the core foils oriented perpendicular to the thin direction of the face sheets.) Honeycomb structures are used in aircraft applications. They are designed to be very light weight with high specific strength and stiffness. (ACT)

**hydrate.** An aluminum oxide with three molecules of chemically combined water. (Dirt)

**impact extrusion.** The process starts with the blanking of slugs from aluminum sheet or plate. The slugs are placed one at a time in the cup-shaped die of a power press. A mechanically operated punch, the diameter of which is equal to the inside diameter of the tube, container, or other object to be formed, strikes the slug a single downward blow with tremendous force. The diameter of the die is equal to the outside diameter of the object to be formed, and the blow of the punch causes the metal to "squirt" up around the punch through the clearance between the punch and the die. The base of the punch is of the same shape as the inside of the top or closed end of the object being extruded. Thus the same blow

that forms the sides of the objects also forms its base, or in the case of collapsible tubes, the shoulder and neck. When the punch is withdrawn, the tube or container is stripped off and another slug placed in the die cavity ready for the next downward stroke of the punch. Best known of the applications for aluminum impact extruded products are the collapsible tubes employed in peacetime in the packaging of toothpastes, shaving creams, ointments, cements, paints, inks, and cleaners. Hooker tubing, made by the impact extrusion process, is used for the shells of flashlights and mechanical pencils. (ANL, December 1943: 6)

**ingot.** A cast form suitable for remelting or fabricating. An ingot may take many forms: some may be 30 feet long and weigh 15 tons; others are notched or specially shaped for stacking and handling. (Dirt)

**interlocking.** An assembly technique for aluminum profiles. Its advantage is that interlocking aluminum profiles can be directly incorporated into an assembly of elements without the use of screws, etc. Interlocking elements have all the advantages of rigid joinery with the possibility of mobility. (EA)

**London Metal Exchange (LME).** The international trading body that facilitates the worldwide open market buying and selling of metals. (Dirt)

**magnesium.** A light, silvery, moderately hard, metallic element used in processing metals and chemicals, and alloying aluminum to give it desired metallurgical properties. (Dirt)

**mill product.** Metal that has been fabricated into an intermediate form before being made into a finished product. The most common fabricating processes for aluminum are rolling, extruding, forging, and casting. Example: aluminum sheet, a mill product, is used to make beverage cans, a finished product. (Dirt)

**permanent mold cast aluminum.** It is made by pouring molten metal into permanent steel molds by hand. The metal is allowed to spread to all parts of the mold by gravity. Excess air escapes through vents leaving a smooth surface with no rough spots or crevices. Pots and pans made by this method are sturdy, heavy, and the aluminum may be graduated (heavy bottoms, thinner sides and covers to transmit heat uniformly). They will last a lifetime. They are of one-piece construction with no rivets or joints to come apart or collect dirt. Cast aluminum is ideal for waterless cooking because its thickness increases the natural heat-retaining qualities of aluminum. Cast aluminum is better adapted for the top of stove cooking than oven cooking. (Dobell)

**pot.** In aluminum production, the electrolytic reduction cell, commonly called a "pot," in which alumina dissolved in molten cryolite is reduced to metallic aluminum. A series of cells connected electrically is called a potline. (Dirt)

**pre-lacquering.** This process consists of preparing the surface while applying a lacquer (usually liquid) to the spools of aluminum. Lacquering is carried out before the final shaping of profiles, and therefore must be resistant to any distortion. (EA)

**rolling.** Ingot is hot-rolled down to plate or slab. The slab is subsequently cold-rolled to sheet or foil thickness, with immediate annealing where necessary. (TALAT)

**smelt.** To fuse or melt ore in order to extract or refine the metal it contains. (Dirt)

**soldering.** An assembly technique that joins aluminum sheets and profiles, creating a construction of great overall structural strength with a minimum of weight. Soldering is primarily used in aeronautics, shipbuilding, and transportation. (EA)

**shell-fusion.** See Hitchner vacuum sand-casting process.

**spinning of aluminum.** It is a process in which aluminum blanks are spun on a lathe by a skilled spinning-lathe operator. The latter, through his ability to "feel" the metal as it is shaped, senses the right amount of pressure to be put against the tool to avoid overdrawing or overstressing the aluminum blank and thus creates uniformly strong spinnings. An inexpensive hardwood block dictates the shape of the spun object. Most symmetrical articles having a circular cross-section normal to the axis of rotation can be produced in small quantities easily and at low cost by spinning. Costly tooling and die sinking are avoided, and the manufacturing cost per unit is quite low when compared with other methods of production. Dimensions can be controlled to reasonably close tolerances, and many semi-precision parts for the aircraft industry are made by this method. Spinning is one of the oldest metalworking arts but modern technology has vastly increased its usefulness to the automobile, aircraft, machinery, and ornamental metalworking industries. Spinning provides a low-cost method of bridging the gap between hand-hammered items and production-line drawing or stamping. Spun aluminum pieces include cooking utensils, lighting reflectors, processing kettles, and ornamental objects. Developments of technique and aluminum alloys have made possible the spinning of articles up to ninety inches in diameter and the use of sheet up to one-half inch in thickness. The strength and heat conductivity of aluminum combine with ease of spinning to make aluminum an ideal metal for this type of manufacture. (ANL, July 1940: 5)

**spun aluminum.** This is a term used to identify the scratched surface of certain pressed aluminum utensils and serving pieces. The outsides of most aluminum pieces are polished to a silver finish. However, an emery cloth is used to scratch the outside surface of spun finished aluminum. Spun aluminum is used quite often for serving pieces that go to the table along with the guests. Example, bun or biscuit warmers, salad bowls, and ice tubs. (Dobell)

**stamped aluminum.** This is made by running sheets of cold aluminum through a stamping machine. In one quick operation an almost finished utensil is turned out. This type of aluminum is so thin it is less durable. However, because less metal is used and because fewer operations in manufacturing are required the price is low. Food preparation gadgets such as cookie cutters, funnels, and molds as well as pots and pans are made from stamped aluminum. (Dobell)

**superplastic aluminum alloys.** Of recent interest for certain applications are the forming (stretching) of "superplastic" alloys of aluminum, for example, aluminum with 6 percent copper and 0.5 percent zirconium. These alloys can be elongated to more than ten times their original length if their grain size is small enough and high temperatures and low stress rate are applied. Advantages to the prolonged production process are high precision, even for the most complicated forms, and low tooling costs. (AL, s.v. "deep drawing", trans. from "Tiefziehen")

**surface treatment.** This process enriches the decorative aspects of the metal and simultaneously protects against corrosion. For definitions of specific surface treatments, see *anodizing*, *pre-lacquering*, and *thermo-lacquering*. (EA)

**thermo-lacquering.** Thermo-lacquering colors the aluminum and offers an infinite color scheme range. It is carried out at the halfway stage in the shaping of profiles, and consists of preparing the metal surface, followed by a spray of paint. (EA)

**vacuum forming.** A manufacturing process by which a sheet of plastic substrate (ABS, styrene, PETG, etc.) is heated until it is softened and is then "stretched" over a custom-built mold. Vacuum is then applied to "pull" all the excess air from the space between the mold and the plastic until they meet, thus forming the part. (FPI)

**white metals.** Babbitt metal or antifriction metal was first produced by Isaac Babbitt in 1839. In present-day usage the term is applied to a whole class of silver-white bearing metals, or "white metals." These alloys usually consist of relatively hard crystals embedded in a softer matrix, a structure important for machine bearings. They are composed primarily of tin, copper, and antimony, with traces of other metals added in some cases and lead substituted for tin in others. (NCE)

**wrought.** Product has been shaped by mechanical working, as by rolling, extruding, or forging. (EPST)

## Aluminum versus Aluminium

The spelling of the word *aluminum* was a matter of debate in the United States at the end of the 19th century. In a lecture delivered at the Franklin Institute in January 1891, Joseph W. Richards, a noted authority on aluminum, discussed the issue. "You all know that two ways of spelling [aluminum] are in common use, both Webster and Worcester sanctioning either way, Webster giving *aluminum* as preferable, and Worcester *aluminium*." Richards preferred the additional "i," which, according to him, was accepted by the worldwide scientific community. In America, however, *aluminum* was popular. "They say it is shorter (true) and *ergo* the better way. That is what I should call intensely American. Mr. Eugene H. Cowles tells us that the Cowles Electric Smelting and Aluminum Company was originally organized as the Cowles Electric Smelting and Alumin(i)um Company but that the writing of the extra *i*, and especially the dotting of it became such an insufferable burden and expense that they went to the court about it and had the name changed to the Cowles Electric Smelting and *Aluminum* Company."

In the minutes of the first meeting of the Pittsburgh Reduction Company in July 1888, both spellings are used. The minutes of an October 30, 1889, meeting included the following: "Resolved that the name of our product be laid over for one week to decide which we shall adopt aluminum or aluminium." The minutes of later meetings contain no evidence of a formal decision, but the company obviously adopted the shorter spelling. It seems that the anecdotal "printer's error" never occurred and the Pittsburgh Reduction Company's spelling of *aluminum*, along with that of the Cowles company, was a conscious choice that helped determine the eventual normal spelling in North America, although the justification for dropping the second "i" is still debatable.

**Sources:** George B. Kauffman, "Aluminum: Its Discovery, Properties, Uses, and Production," *Cahiers d'Histoire de l'Aluminium* 5 (Autumn 1989): 41–42; Joseph W. Richards, "The Aluminium Problem," *Journal of the Franklin Institute* (March 1891): 189–91.

## Further Reading

### Books

Adam, Peter. *Eileen Gray: Architect/Designer.* New York: Abrams, 1987.

Antonelli, Paola. *Mutant Materials in Contemporary Design.* New York: Museum of Modern Art, 1995.

*Austerity to Affluence: British Art & Design, 1945–1962.* London: Merrell Holberton, 1997.

Barten, Sigrid. *René Lalique: Schmuck u. Objets d'art, 1890–1910.* Munich: Prestel, 1977.

Brubaker, Sterling. *Trends in the World Aluminum Industry.* Baltimore: Johns Hopkins University Press, 1967.

Bunten, Elaine D., John L. Donaldson, and Eugene C. McDowell. *Hazard Assessment of Aluminum Electrical Wiring in Residential Use.* Washington, D.C.: National Bureau of Standards, 1974.

Burkhart, Bryan, and David Hunt. *Airstream: The History of the Land Yacht.* San Francisco: Chronicle Books, 2000.

Bush, Donald J. *The Streamlined Decade.* New York: George Braziller, 1975.

Campbell, Bonita J. *Wendell August Forge: Seventy Five Years of Artistry in Metal.* Upland, Calif.: Dragonflyer Press, 1999.

Carr, Charles C. *ALCOA: An American Enterprise.* New York: Rinehart & Co., 1952.

Corn, Joseph J. *The Winged Gospel: America's Romance with Aviation, 1900–1950.* New York: Oxford University Press, 1983.

Cowan, Ruth Schwartz. *More Work for Mother: The Ironies of Household Technology from the Open Hearth to the Microwave.* New York: Basic Books, 1983.

Deville, Henri Sainte-Claire. *De l'aluminium: Ses propriétés, sa fabrication et ses applications.* Paris: Mallet-Bachelier, 1859.

Doordan, Dennis P. "Promoting Aluminum: Designers and the American Aluminum Industry." In *Design History: An Anthology,* edited by Dennis P. Doordan. Cambridge: MIT Press, 1995.

Dorment, Richard, and Isabelle Anscombe. *Eros by Sir Alfred Gilbert.* London: Fine Art Society, 1987.

Edwards, Junius. *The Immortal Woodshed.* New York: Dodd, Mead, 1955.

Fenichell, Stephen. *Plastic: The Making of a Synthetic Century.* New York: HarperBusiness, 1996.

Fiell, Charlotte, and Peter Fiell. *1000 Chairs.* Cologne: Taschen, 1997.

Frederick, Christine. *The New Housekeeping: Efficiency Studies in Home Management.* Garden City, N.Y.: Doubleday, Page, 1913.

Friedel, Robert. "Some Matter of Substance." In *History from Things: Essays on Material Culture,* edited by Steven Lubar and W. David Kingery. Washington, D.C.: Smithsonian Institution Press, 1993.

Gignoux, Claude J. *Histoire d'une entreprise française.* Paris: Hachette, 1955.

Goldberg, Vicki. *Margaret Bourke-White: A Biography.* Reading, Mass.: Addison-Wesley, 1987.

Gordon, J. E. *Structures: Or, Why Things Don't Fall Down.* New York: Plenum Press, 1978.

Graham, Margaret B. W., and Bettye H. Pruitt. *R&D for Industry: A Century of Technical Innovations at Alcoa.* Cambridge: Cambridge University Press, 1990.

Greif, Martin. *Depression Modern: The Thirties Style in America.* New York: Universe Books, 1975.

Grinberg, Ivan, Pascal Griset, and Muriel Le Roux, eds. *Cent ans d'innovation dans l'industrie aluminium.* Paris: Editions l'Harmattan, 1997.

Guidot, Raymond, and Alain Guiheux. *Jean Prouvé, Constructeur.* Paris: Centre Georges Pompidou, 1990.

Guidot, Raymond, and Oliver Boissière. *Ron Arad.* Paris: Editions Dis Voir, 1997.

Hachez-Leroy, Florence. *L'aluminium français: L'invention d'un marché, 1911–1983.* Paris: CNRS, 1999.

Hanks, David A., and Jennifer Toher. *Donald Deskey: Decorative Designs and Interiors.* New York: Dutton, 1987.

Hine, Thomas. *The Total Package: The Evolution and Secret Meanings of Boxes, Bottles, Cans, and Tubes.* Boston: Little, Brown, 1995.

*Issey Miyake: Making Things.* Paris and Zurich: Fondation Cartier pour l'Art Contemporain and Scalo, 1998.

Joliet, Hans, ed. *Aluminium: Die ersten hundert Jahre.* Düsseldorf: VDI Verlag, 1988.

Kamitsis, Lydia. *Paco Rabanne: A Feeling for Research.* Paris: Editions M. Lafon, 1996.

Kaplan, Wendy, ed. *Designing Modernity: The Arts of Reform and Persuasion, 1885–1945.* New York: Thames and Hudson, 1995.

Kirkham, Pat. *Charles and Ray Eames: Designers of the Twentieth Century.* Cambridge: MIT Press, 1995.

Krausse, Joachim, and Claude Lichtenstein, eds. *Your Private Sky: R. Buckminster Fuller. The Art of Design Science.* Baden: Lars Müller, 1999.

Küper, Marijke, and Ida van Zijl. *Gerrit Th. Rietveld.* Utrecht: Centraal Museum, 1992.

Lifshey, Earl. *The Housewares Story: A History of the American Housewares Industry.* Chicago: National Housewares Manufacturers Association, 1973.

Lipman, Jonathan. *Frank Lloyd Wright and the Johnson Wax Buildings.* New York: Rizzoli, 1986.

Manzini, Ezio. *The Material of Invention: Materials and Design.* Cambridge: MIT Press, 1989.

Marcilhac, Félix. *René Lalique, maître-verrier, 1860–1945: Analyse de l'œuvre et catalogue raisonné.* New ed. Paris: Editions de l'Amateur, 1994.

McCarty, Cara, and Matilda McQuaid, eds. *Structure and Surface: Contemporary Japanese Textiles.* New York: Museum of Modern Art, 1998.

Meikle, Jeffrey L. *American Plastic: A Cultural History.* New Brunswick, N.J.: Rutgers University Press, 1995.

Morel, Paul, ed. *Histoire technique de la production d'aluminium.* 2d ed. Grenoble: Presses Universitaires de Grenoble, 1992.

Mumford, Lewis. *Technics and Civilization.* New York: Harcourt, Brace, 1934.

Neuhart, John, Marilyn Neuhart, and Ray Eames. *Eames Design: The Work of the Office of Charles and Ray Eames*. New York: Abrams, 1989.

Noblet, Jocelyn de, ed. *Industrial Design: Reflection of a Century*. Paris: Flammarion, 1993.

Norman, Donald A. *The Design of Everyday Things*. New York: Doubleday, 1989.

Nye, David E. *American Technological Sublime*. Cambridge: MIT Press, 1994.

Nye, David E. *Electrifying America: Social Meanings of a New Technology, 1880–1940*. Cambridge: MIT Press, 1990.

Ogden, Annegret S. *The Great American Housewife: From Helpmate to Wage Earner, 1776–1986*. Westport, Conn.: Greenwood Press, 1986.

Osterreichische Postsparkasse. *Otto Wagner: The Austrian Postal Savings Bank*. Vienna: Falter-Verlag, 1996.

Papanek, Victor. *The Green Imperative: Natural Design for the Real World*. New York: Thames and Hudson, 1995.

Patton, Phil. *Made in USA: The Secret History of the Things That Made America*. New York: Grove Weidenfeld, 1992.

Peck, Merton J. *Competition in the Aluminium Industry, 1945–1958*. Cambridge: Harvard University Press, 1961.

Pendergrast, Mark. *For God, Country, and Coca-Cola: The Unauthorized History of the Great American Soft Drink and the Company That Makes It*. New York: Macmillan, 1993.

Petroski, Henry. *The Evolution of Useful Things*. New York: Knopf, 1992.

Petroski, Henry. *Invention By Design: How Engineers Get from Thought to Thing*. Cambridge: Harvard University Press, 1996.

Phillips, Lisa. *Frederick Kiesler*. New York: Whitney Museum of American Art, 1989.

Pulos, Arthur. *The American Design Adventure, 1940–1975*. Cambridge: MIT Press, 1988.

Richardson, Brenda. *Scott Burton*. Baltimore: Baltimore Museum of Art, 1986.

Richards, Joseph W. *Aluminium: Its History, Occurrence, Properties, Metallurgy and Applications, Including Its Alloys*. 3d ed. Philadelphia: H. C. Baird & Co., 1896.

Schäfke, Werner, Thomas Schleper, and Max Tauch, eds. *Aluminium: Das Metall der Moderne: Gestalt, Gebrauch, Geschichte*. Cologne: Kölnisches Stadtmuseum, 1991.

Schatzberg, Eric. *Wings of Wood, Wings of Metal: Culture and Technical Choice in American Airplane Materials, 1914–1945*. Princeton: Princeton University Press, 1999.

Schön, Donald A. *Technology and Change: The New Heraclitus*. Oxford: Pergamon, 1967.

*The Second Empire, 1852–1870: Art in France under Napoleon III*. Philadelphia: Philadelphia Museum of Art, 1978.

Smith, George David. *From Monopoly to Competition: The Transformation of Alcoa, 1888–1986*. Cambridge: Cambridge University Press, 1988.

Sparke, Penny. "'From a Lipstick to a Steamship': The Growth of the American Design Profession." In *Design History: Fad or Function?* London: Design Council, 1978.

Sparke, Penny, ed. *The Plastics Age: From Modernity to Post-Modernity*. London: Victoria and Albert Museum, 1990.

Stern, Jewel. "Striking the Modern Note in Metal." In *Craft in the Machine Age, 1920–1945*, edited by Janet Kardon. New York: Abrams, 1995.

Sudjic, Deyan. *Ron Arad*. London: Laurence King Publishing, 1999.

Sulzer, Peter. *Jean Prouvé: Complete Works*, vol. 2, *1934–1944*. Basel: Birkhauser, 2000.

Treib, Marc, and Dorothée Imbert. *Garrett Eckbo: Modern Landscapes for Living*. Berkeley: University of California Press, 1997.

Varnedoe, Kirk. *Vienna 1900: Art, Architecture & Design*. New York: Museum of Modern Art, 1986.

Vegesack, Alexander von, Peter Dunas, and Mathias Schwartz-Clauss, eds. *100 Masterpieces from the Vitra Design Museum Collection*. Weil am Rhein, Germany: Vitra Design Museum, 1996.

Wallace, Donald H. *Market Control in the Aluminum Industry*. Cambridge: Harvard University Press, 1937.

Westerman, Nada, and Joan Wessel. *American Design Classics*. New York: Design Publication, Inc., 1985.

Wilk, Christopher. *Marcel Breuer: Furniture and Interiors*. New York: Museum of Modern Art, 1981.

Wilson, Richard Guy, Dianne H. Pilgrim, and Dickran Tashjian. *The Machine Age in America, 1918–1941*. New York: Abrams, 1986.

## Periodicals

"Adventure in Aluminum: Russel Wright was one of the first designers to explore its potential." *Industrial Design* 7, no. 5 (May 1960): 46–53.

"Aluminum—A Basic Material of Tomorrow." *Interiors* 103, no. 11 (June 1944): 65–70+.

"Aluminum Reborn." *Fortune* 33, no. 5 (May 1946): 102–09+.

Binczewski, George J. "The Point of the Monument: A History of the Aluminum Cap of the Washington Monument." *JOM* 47, no. 11 (November 1995): 20–25.

Bonney, Louise. "New Metal Furniture for Modern Schemes." *House & Garden* 57, no. 4 (April 1930): 82–84+.

Canby, Thomas Y. "Aluminum, the Magic Metal." *National Geographic* 154, no. 2 (1978): 186–211.

"Chairs of Aluminum on a Quantity Production Basis in a Buffalo Factory." *Furniture Manufacturer* 37, no. 5 (May 1929): 60–64+.

Craig, Norman C. "Charles Martin Hall—The Young Man, His Mentor, and His Metal." *Journal of Chemical Education* 63, no. 7 (July 1986): 557–59.

Giovannini, Joseph. "Seat of Authority." *I.D. Magazine* 46, no. 3 (May 1999): 44–47.

Hobbs, Douglass B. "Aluminum—A Decorative Metal." *Good Furniture and Decoration* 35, no. 2 (August 1930): 89–93.

Maier, Helmutt. "New Age Metal or Ersatz? Technological Uncertainties and Ideological Implications of Aluminium up to the 1930s." *ICON* 3 (1997): 181–201.

Plateau, Jean, and Elise Picard. "Le fléau de la balance ou les apparitions de l'aluminium: L'Exposition de 1855." *Cahiers d'Histoire de l'Aluminium* 23 (Winter 1998): 9–28.

Smith, Gregory W. "Alcoa's Aluminum Furniture: New Applications for a Modern Material, 1924–1934." *Pittsburgh History* 78, no. 2 (Summer 1995): 52–64.

Sonderegger, Christina. "Aluminiumzeit." *Kunst+Architektur in der Schweiz* 48, no. 3 (March 1997): 20–29.

Dimensions are in inches and centimeters. Except where noted, height precedes width precedes depth.

## Introduction

**1.1** Bracelet (French), ca. 1858, aluminum and gold, 1¼ x 2½ x 2¼ in. (3.1 x 6.3 x 5.7 cm); Carnegie Museum of Art, Martha Mack Lewis Fund, 1998.5

**1.2** Beads (African), 20th century, aluminum, L. 55 in. (139.7 cm); Lent from a private collection

**1.3** Marcia Lewis (American, b. 1946) Creature Collar II, ca. 1978–80, aluminum, 3 x 10½ x 9 in. (7.6 x 26.6 x 22.8 cm); Carnegie Museum of Art, Gift of the artist, 1999.39.4

**1.4** Pathé (French, ca. 1895–1928) Phonograph, ca. 1906–08, aluminum, oak, nickel-plated steel, and metal, 30¼ x 25½ x 21¾ in. (77 x 65 x 55 cm); Lent by Jean Plateau—Institut pour l'Histoire de l'Aluminium Collection, Paris; Marked: Les disques Pathé chantent sans aiguille / marque de fabrique déposée / Reproducteur pour disques Pathé, breveté s.g.d.g. / Pathé concert

**1.5** John Vassos, designer (American, b. Romania, 1898–1985); Radio Company of America, now owned by Thomson Multimedia, manufacturer (American, 1919–present) RCA Victor Special, model N, ca. 1937, aluminum, chrome-plated steel, velvet, and plastic, 8 x 17½ x 17½ in. (20.3 x 44.5 x 43.8 cm); Lent by Mitchell Wolfson Jr. Collection, The Wolfsonian—Florida International University, Miami Beach

**1.6** Studio Naço (French, 1987–present) Bianca M radio, 1989, polished aluminum and other materials, 12⅝ x 5¼ x 12⅝ in. (32 x 13.5 x 32 cm); Lent by Centre du Georges Pompidou, Paris, Musée National d'Art Moderne / Centre de Création Industrielle MNAM / CCI

**1.7** Sony Electronics Inc. (Japanese, 1946–present) Walkman, 1979, aluminum and other materials, 4½ x 3 x ¾ in. (11.4 x 7.6 x 1.9 cm); Lent by Sony Electronics Inc.

**1.8** Bang & Olufsen (Danish, 1925–present) Five BeoLab 4000 loudspeakers, 2000; designed 1997, manufactured 1997–present; anodized aluminum and other materials, 12 x 11 x 5⅛ in. (30 x 27.9 x 13 cm); Lent by Bang & Olufsen

## Inventing Aluminum

**2.1** Friedrich Wöhler (German, 1800–1882) Test tube with aluminum particles, 1845, aluminum and glass, L. 3⅞ in. (10 cm); Lent by Deutsches Museum, Munich

**2.2** Collot Frères, later renamed Collot-Longue et Cie (French, 1848–1920) Balance arm, 1855, set in later scale; balance arm: aluminum; scale: aluminum, brass, and agate; balance arm: L. 11 x H. 7⅞ in. (28 x 20 cm); overall: 21⅞ x 25¾ x 16½ in. (55.5 x 65.5 x 42 cm); Lent by Musée des Arts et Métiers—CNAM—Paris; Inscription on arm: Collot Fres / exposition 1855

**2.3** Gabriel Jules Thomas, designer (French, 1824–1905); F. Barbedienne Foundry, maker (French, 1839–1955) Bust of Henri Sainte-Claire Deville, designed 1882, cast by 1900; cast aluminum, 9¾ x 5½ x 14⅝ in. (25 x 14 x 37 cm); Lent by Deutsches Museum, Munich; Inscription: Henri Sainte-Claire Deville / Membre de l'Institut / G. J. Thomas 1882; Signed: F. Barbedienne, Fondeur

**2.4** Charles Rambert, designer (French, active Paris after ca. 1848); Honoré-Séverin Bourdoncle, maker (French, 1823–1893) Baby rattle for the Prince Imperial, 1856, aluminum, gold, coral, emeralds, and diamonds, L. 8⅛ x w. 1⅞ in. (20.5 x 4.7 cm); Lent by S.A.I. La Princesse Napoléon

**2.5** Charles Christofle, designer (French, 1805–1863); Charles Christofle et Cie, manufacturer (French, 1830–present) Centerpiece, 1858, aluminum, silvered-copper alloy, and gilded bronze, 13¾ x 15¾ x 12 in. (35 x 40 x 30.5 cm); Lent by Musée National de Compiègne; Inscription translated from Latin: "To the Emperor Napoléon III for the help and encouragement that he has brought to the dedicated work of the learned Henri Sainte-Claire Deville in the making of aluminum—the goldsmith Charles Christofle offered respectfully the statues made in the new metal"

**2.6** F. Desbœufs, engraver (French, active mid-19th century) Medal in original box, 1856, aluminum, wood, morocco leather, velvet, silk, and gold leaf, DIAM. 1⅜ in. (3.5 cm); Lent by Jean Plateau—Institut pour l'Histoire de l'Aluminium Collection, Paris; Marked: Napoléon III Empereur; Engraved: La Société de Javél à Mr Malpas-Duché 1856; Box marked: ALUMINIUM

**2.7** Jean-Auguste Barre (French, 1811–1896) Eagle, designed 1860, gilded aluminum, H. 10¾ x w. 7⅞ in. (27.5 x 20 cm); Lent by Musée de l'Armée, Paris

**2.8** Jørgen Balthasar Dalhoff (Danish, 1800–1890) Parade helmet for The Hereditary Prince Ferdinand, ca. 1859, aluminum and other materials, 12½ x 12¼ x 7⅜ in. (31.7 x 30.6 x 18.8 cm); Lent by The Royal Danish Collections, Rosenborg Castle, Copenhagen

**2.9** Gustave Ponton d'Amécourt (French, 1825–1888) Model of a steam helicopter, 1863, aluminum, brass, and copper, H. 25⅝ x DIAM. 19⅝ in. (65 x 50 cm); Lent by Musée de l'Air et de l'Espace

**2.10** Charles Christofle, designer (French, 1805–1863); Charles Christofle et Cie, manufacturer (French, 1830–present) Stand, 1858, chased and gilded aluminum, H. 9¼ x DIAM. 9¾ in. (23.4 x 24.8 cm); Lent by Musée d'Orsay, Paris (on loan from Musée de L'Impression sur Etoffes et du Papier Peint à Mulhouse)

**2.11** Stand (probably French), ca. 1875, cast and chased aluminum, 5⅛ x 8⅝ x 8⅝ in. (13 x 22 x 22 cm); Lent by Jean Plateau—Institut pour l'Histoire de l'Aluminium Collection, Paris

**2.12** Stand (probably French), ca. 1860, chemically engraved aluminum, 5⅝ x 6 x 6 in. (14.4 x 15.1 x 15.1 cm); Lent by Jean Plateau—Institut pour l'Histoire de l'Aluminium Collection, Paris

**2.13** Arthur Martin, designer (French); Paris Mint, manufacturer (French, 1795–present) Medallion given at birth, before 1860, aluminum, DIAM. 2 in. (5 cm); Lent by Jean Plateau—Institut pour l'Histoire de l'Aluminium Collection, Paris; Marked: Je renonce à Satan, à ses pompes et à ses oeuvres / Baptême [—] 18 [—] / et Confirmation [—] 18 [—] / Celui qui mange ma chair et boit mon sang a la vie éternelle / Prem. Communion [—] 18 [—] / Ste. Barbe; on edge: ALUMINIUM

**2.14** Albert Barre, general engraver for Paris Mint (French, active 1855–78); Paris Mint, manufacturer (French, 1795–present) Medallion, ca. 1858, aluminum, DIAM. 2⅜ in. (5.9 cm); Lent by Jean Plateau—Institut pour l'Histoire de l'Aluminium Collection, Paris; Marked: Louis Jacques Thénard / La Sociéte à la mémoire de son fondateur, 1858 / Société de secours des amis des sciences fondée le mars 1857

**2.15** Charles Emile Matthis, author and illustrator (French, 1838–unknown); Jouvet et Cie, printer (French, active second half of 19th century) Les Deux Gaspards, 1887, board, paper, aluminum leaf, gold leaf, and linen, 10⅜ x 8¹⁄₁₆ x 1 in. (26.2 x 20.5 x 2.5 cm); Lent by Jean Plateau—Institut pour l'Histoire de l'Aluminium Collection, Paris

**2.16** Morizot, printer (French, active second half of 19th century)
Prayer book, ca. 1870, chased aluminum, aluminum leaf, board, paper, electroplated silver, and velvet, 5⅛ x 3¾ x 1⅛ in. (12.8 x 9.5 x 2.8 cm); Lent by Jean Plateau—Institut pour l'Histoire de l'Aluminium Collection, Paris

**2.17** Opera glasses (French), ca. 1865, aluminum and glass, 2⅜ x 5⅜ x 1⅝ in. (6 x 13.5 x 4 cm); Lent by Jean Plateau—Institut pour l'Histoire de l'Aluminium Collection, Paris

**2.18** Opera glasses (French), ca. 1875, aluminum, glass, nickel-plated metal, and mother-of-pearl, 2⅜ x 4⅜ x 1⅝ in. (6 x 11 x 4.3 cm); Lent by Jean Plateau—Institut pour l'Histoire de l'Aluminium Collection, Paris; Marked: HD / Paris

**2.19** August Edouard Achille Luce, designer; Crespo De Borbon, retailer (Cuban)
Fan, ca. 1866, aluminum, silk, and paper, 11¼ x 20⅝ x ¾ in. (28.5 x 52.5 x 2 cm); Lent by National Museum of American History, Smithsonian Institution; Inscription: Barte Crespo De Borbon… the patent plated fan…. Habana—Paris

**2.20** Suite of jewelry: brooch, earrings, and cuff studs (probably French), ca. 1860, aluminum and gold; brooch: 2¼ x 1⅞ x ½ in. (5.8 x 4.8 x 1.3 cm); earrings: 2½ x ¾ x 7/16 in. (6.4 x 1.9 x 1.1 cm) and 2⅝ x ¾ x ⅜ in. (6.5 x 1.9 x 1.0 cm); cuff studs: DIAM. ⅞ x D. ⅝ in. (2.2 x 1.5 cm) and DIAM. ⅞ x D. ⅝ in. (2.2 x 1.5 cm); Cooper-Hewitt National Design Museum, Smithsonian Institution / Art Resource, NY, Gift of Sarah Cooper Hewitt, Eleanor Garnier Hewitt, and Mrs. James O'Green, in memory of their father and mother, Mr. and Mrs. Abram S. Hewitt, 1928–5–3a,b, and Gift of Mrs. Gustav E. Kissel, 1928–5–4a/c

**2.21** Attributed to Honoré Séverin-Bourdoncle (French, 1823–1893)
Bracelet, ca. 1858, aluminum and gold, H. 2⅝ x DIAM. 5⅜ in. (6.6 x 13.5 cm); Lent by Musée des Arts Décoratifs, Paris

**2.22** Pansy brooch (French), ca. 1865, aluminum, vermeil, and silver, 1⅞ x 1⅜ x ⅜ in. (4.6 x 3.5 x 1 cm); Lent by Jean Plateau—Institut pour l'Histoire de l'Aluminium Collection, Paris

**2.23** Flower brooch (French), ca. 1870, aluminum and gilded metal, 1½ x 1½ x ⅝ in. (3.8 x 3.8 x 1.6 cm); Lent by Jean Plateau—Institut pour l'Histoire de l'Aluminium Collection, Paris

**2.24** Cameo brooch (French), ca. 1870, aluminum, gilded metal, and carved shell, 1¼ x 1¾ x ⅜ in. (3.2 x 4.3 x 0.9 cm); Lent by Jean Plateau—Institut pour l'Histoire de l'Aluminium Collection, Paris

**2.25** Victor Chapron (French, active 1860–67)
Bracelet, ca. 1865, embossed and chased aluminum, gilded metal, and garnets, ⅜ x 7¾ x 1⅛ in. (1 x 19.8 x 2.9 cm); Lent by Jean Plateau—Institut pour l'Histoire de l'Aluminium Collection, Paris

**2.26** Attributed to Frédéric Milisch (French, active 1864–72)
Bracelet, ca. 1865, embossed and chased aluminum and vermeil, 6¾ x ⅞ x ⅜ in. (17.2 x 2.1 x 0.9 cm); Lent by Jean Plateau—Institut pour l'Histoire de l'Aluminium Collection, Paris

**2.27** Armand Dufet (French, active 1849–70)
Bracelet, ca. 1860, chased aluminum and gold, ¾ x 3⅛ x 5½ in. (1.9 x 8 x 14 cm); Lent by Jean Plateau—Institut pour l'Histoire de l'Aluminium Collection, Paris

**2.28** Perfume bottle in original case (probably French), ca. 1860, aluminum, glass, gold, and leather, 2⅜ x 1⅝ x 5½ in. (6 x 4 x 14 cm); Lent by Alusuisse Technology & Management Ltd., Communications TCA, Neuhausen, Switzerland

**2.29** Société Paul Morin et Cie, later renamed Société Anonyme de l'Aluminium, manufacturer (French, 1857–85); Maison de l'Aluminium, retailer (French, 1875–91)
Six spoons in original box, ca. 1880, aluminum bronze, wood, leather, and gold leaf, each 4⅐ x ¾ x ½ in. (10.5 x 1.9 x 1.3 cm); Lent by Jean Plateau—Institut pour l'Histoire de l'Aluminium Collection, Paris; Box marked on inside: Médaille d'or, Maison de l'ALUMINIUM / Victor TESTEVUIDE / Paris, 21 Bould Poissonnière, 21; spoons marked: P Morin 10

**2.30** Cruet stand (probably French), ca. 1875, aluminum, metal, and cut glass, 10½ x 13⅞ x 6⅔ in. (26.7 x 35 x 17 cm); Lent by Jean Plateau—Institut pour l'Histoire de l'Aluminium Collection, Paris

**2.31** Set of six napkin rings in original box (possibly British), ca. 1870, aluminum, leather, wood, velvet, and other materials; each W. 1¹³/₁₆ x DIAM. 1⁵/₁₆ in. (4.6 x 3.3 cm); Carnegie Museum of Art, Purchase: Gifts in memory of Harry Blum and the Helen Johnston Acquisition Fund, 1998.58

**2.32** Colonel Thos. Lincoln Casey, designer (American, 1831–1896); Eastern Nonferrous Foundry, originally William Frishmuth's foundry, maker (American, ca. 1855–present)
Cap for Washington Monument, 1984, exact replica of the 1884 original; aluminum, H. 9 x W. 5½ in. (22.9 x 14 cm); Lent by Oberlin College and George J. Binczewski; Front inscription: Joint Commission at Setting of Capstone, Chester A. Arthur, W. W. Corcoran, Chairman, M. E. Bell, Edward Clark, John Newton; right side inscription: Corner Stone laid on bed of foundation, July 4, 1848; First stone at Height of 152 feet laid, August 7, 1880; Capstone set December 6, 1884; rear inscription: Chief Engineer and Architect, Thos. Lincoln Casey, Colonel, Corps of Engineers, Assistants, George W. Davis, Captain, 14th Infantry; Bernard R. Green, Civil Engineer; Master Mechanic, P. H. McLaughlin; left side inscription: LAUS DEO

**2.33** Charles Martin Hall (American, 1863–1914)
Globules of aluminum, 1886–88, aluminum; largest: H. 1 x DIAM. 4 in. (2.5 x 10.2 cm); Lent by Alcoa Inc.

**2.34** Louis Blum (American, b. 1909)
Box holding Charles Martin Hall's globules of aluminum, ca. 1935, aluminum, 5¾ x 14 x 8½ in. (14.6 x 35.6 x 21.6 cm); Lent by Alcoa Inc.

**2.35** Pittsburgh Reduction Company, later renamed Alcoa (American, 1888–present)
Price & Fact Sheet, ca. 1894, aluminum and paint, 10¾ x 8 in. (27 x 20.3 cm); Lent by Historical Society of Western Pennsylvania, Gift of Aluminum Company of America

**2.36** Margaret Macdonald (Scottish, 1865–1933); Frances Macdonald (Scottish, 1874–1921)
Picture frame, 1897, aluminum and oak, H. 27³/₁₆ x W. 12¾ in. (69.1 x 32.4 cm); Carnegie Museum of Art, Purchase: Roy A. Hunt Fund, Gift of Hunt Foundation, and Patrons Art Fund, 85.20

**2.37** Talwin Morris (Scottish, 1865–1911)
Buckle, ca. 1900, aluminum set with foil-backed pastes, H. 3 x W. 1⁹/₁₀ in. (7.6 x 4.8 cm); Lent by Mr. John Jesse

**2.38** C. F. A. Voysey (British, 1857–1941)
Clock, ca. 1900, designed 1895; aluminum and copper, H. 20⅞ x D. 6 in. (53 x 15 cm); Lent by The Birkenhead Collection

**2.39** René Lalique (French, 1860–1945)
*The Berenice Tiara*, 1899, aluminum, ivory, and garnets, H. 10⅝ x W. 14⅝ in. (27 x 37 cm); Lent by Musée Lambinet, Versailles

**2.40** Compagnie d'Alais et de la Camargue, manufacturer (French, 1855–1921)
Three spoons, ca. 1900, aluminum bronze, each 5⅝ x 1¼ x ⅝ in. (14.5 x 2.9 x 1.6 cm); Lent by Jean Plateau—Institut pour l'Histoire de l'Aluminium Collection, Paris; Marked: Calypso (trademark)

**2.41** Health Developing Apparatus Co., Inc. (American)
Exercise machine, ca. 1905, aluminum, brass, steel, wood, and cotton rope; folded: 11 x 13 x 24½ in. (27.9 x 33 x 62.2 cm); extended: 8½ x 13 x 49 in. (21.6 x 33 x 124.5 cm); Lent by Mitchell Wolfson Jr. Collection, The Wolfsonian—Florida International University, Miami Beach

**2.42** Aitchison & Co. (British, 1889–present)
Folding binoculars in original case, ca. 1900, aluminum, glass, leather, and metal; fully extended: 3 x 4⅜ x 2¼ in. (7.6 x 11.1 x 5.7 cm); folded: 1⅜ x 5 x 2¹³/₁₆ in. (3.5 x 12.7 x 7.1 cm); Carnegie Museum of Art, Gift of Paul Reeves, 1997.53a–b; Marked: AITCHISON & CO. / Makers / London & Provinces / x6 / 9144½

**2.43** George Blickensderfer, designer (American, d. 1919); Blickensderfer Manufacturing Co., manufacturer (American, active 1893–1915)
Featherweight Blick typewriter, ca. 1894, aluminum, iron, plastic, rubber, copper, and felt, 6½ x 12⅛ x 9¾ in. (16.5 x 30.8 x 24.8 cm); Lent by Mitchell Wolfson Jr. Collection, The Wolfsonian—Florida International University, Miami Beach

2.44 Surgical instruments in original case (probably Swiss), ca. 1890s, aluminum, leather, and other materials, 5⅛ x 16½ x 13⅜ in. (13 x 42 x 34 cm); Lent by Alusuisse Technology & Management Ltd., Communications TCA

2.45 Pocket watch (Swiss), ca. 1890, aluminum and other materials, DIAM. 2 in. (5.3 cm); Lent by Musée Internationale d'Horlogerie, La Chaux-du-Fonds, Switzerland (Inv. 731)

2.46 Pocket watch (Swiss), ca. 1890, aluminum and other materials, DIAM. 2 in. (5.3 cm); Lent by Musée Internationale d'Horlogerie, La Chaux-du-Fonds, Switzerland (Inv. I–623)

2.47 Eastman Kodak Company (American, 1892–present)
8 x 10 flatbed camera with 10-inch wide-field Ektar lens, date unknown, aluminum and other materials; folded: 14 x 12 x 5 in. (35.6 x 30.5 x 12.7 cm); extended: 14 x 12 x 15 in. (35.6 x 30.5 x 38.1 cm); Lent by the Ansel Adams Archive, Center for Creative Photography, Tucson

2.48 Doorplates (probably Swiss), 1890–1900, aluminum, each H. 10⅝ x w. 4¼ in. (27 x 10.5 cm); Lent by Alusuisse Technology & Management Ltd., Communications TCA, Neuhausen, Switzerland

2.49 Shoe (probably Swiss), 1890–1900, aluminum, 3¼ x 2¾ x 8⅝ in. (8 x 7 x 22 cm); Lent by Alusuisse Technology & Management Ltd., Communications TCA, Neuhausen, Switzerland

2.50 Viko, manufacturer (American, first quarter of 20th century)
Set of combs in original box, ca. 1900, aluminum and cardboard; box: 1¼ x 11½ x 8¼ in. (3.2 x 29.2 x 21 cm); Lent by Historical Society of Western Pennsylvania, Gift of Aluminum Company of America

2.51 Door handle (probably Swiss), 1890–1900, aluminum, 2¼ x 6⅜ x 3½ in. (5.5 x 16 x 9 cm); Lent by Alusuisse Technology & Management Ltd., Communications TCA, Neuhausen, Switzerland

2.52 Aluminum Musical Instrument Company (American, active first half of 20th century)
Violin, ca. 1932, natural lacquered aluminum, Bakelite, ebony, metal, and wire, 23½ x 8 x 3½ in. (59.7 x 20.3 x 8.9 cm); Lent by Historical Society of Western Pennsylvania, Gift of Aluminum Company of America

2.53 Archibald Knox, designer (British, 1864–1933); Liberty & Co., retailer (British, 1875–present)
Biscuit box, 1903, pewter, 4⅝ x 4⅝ x 4⅝ in. (11.7 x 11.7 x 11.7 cm); Carnegie Museum of Art, Decorative Arts Purchase Fund, 1999.49.1a–b

2.54 N. C. J., maker (British, first half of 20th century); Carr & Co., retailer (British, ca. 1920–present)
Biscuit box, ca. 1924, aluminum, 4⅛ x 4⅝ x 4⅝ in. (10.5 x 12 x 12 cm); Lent by Jean Plateau—Institut pour l'Histoire de l'Aluminium Collection, Paris

2.55 Two 100 mark notes (German), ca. 1922, aluminum, H. 3³⁄₁₆ x w. 5 in. (8 x 12.7 cm); Carnegie Museum of Art, Anonymous Gift, 1999.8.1–.2

2.56 Quirin Aluminium, later renamed Quiralu (French, 1933–61)
Seven toy rodeo figures, ca. 1945, cast aluminum and paint, H. (maximum) 6¾ in. (17 cm); Lent by Jean Plateau—Institut pour l'Histoire de l'Aluminium Collection, Paris

2.57 New Method Company, manufacturer (American, active ca. 1905)
Tray, ca. 1905, aluminum and ink, ⅜ x 6½ x 4 in. (0.9 x 16.5 x 10.2 cm); Carnegie Museum of Art, Gift of East End Galleries, 1997.32

2.58 Postcard (French), ca. 1906, aluminum sheet, H. 3½ x w. 5½ in. (8.8 x 14 cm); Lent by Jean Plateau—Institut pour l'Histoire de l'Aluminium Collection, Paris; Marked recto: Paris, 15 avril 1906 / Linarès; marked verso: Union Postale Universelle / Carte Postale…

2.59 Postcard (Spanish), ca. 1906, aluminum sheet, H. 5½ x w. 3½ in. (14 x 8.8 cm); Lent by Jean Plateau—Institut pour l'Histoire de l'Aluminium Collection, Paris; Marked recto: Sevilla / Portada del Salacio de San Telmo; marked verso: Tarjeta Postal / Union Postal Universal / ESPANA…

2.60 Two Christmas cards (probably American), ca. 1900, aluminum and paint, 4½ x 3¼ x ¹⁄₁₆ in. (11.4 x 8.3 x 0.2 cm) and 3½ x 2¼ x ¹⁄₁₆ in. (8.9 x 5 x 0.2 cm); Lent by Historical Society of Western Pennsylvania, Gift of Aluminum Company of America

2.61 Set of playing cards (American), ca. 1930, aluminum and paint, 3½ x 2¼ x ¹⁄₁₀ in. (8.9 x 5.7 x 0.3 cm); Lent by Historical Society of Western Pennsylvania, Gift of Aluminum Company of America

2.62 Sir Alfred Gilbert, designer (British, 1854–1934); Compagnie des Bronzes, manufacturer (Belgian, active 1853–1935)
St. George, 1899, cast aluminum, H. 18 in. (45.7 cm); Lent by Trustees of Cecil Higgins Art Gallery, Bedford

2.63 Sir Alfred Gilbert, designer (British, 1854–1934); Cast by George Mancini (b. 1903) at Morris Singer Foundry (British, 1848–present)
Eros, 1985 cast of original (1893) on the Shaftesbury Memorial, Piccadilly Circus, aluminum, 94 x 48 x 60 in. (238.8 x 121.9 x 152.4 cm); Lent by the Fine Arts Society plc, London

2.64 F. Barbedienne Foundry (French, 1839–1955)
Venus de Milo, 1889, cast aluminum, 33⅝ x 10¼ x 8 in. (85.4 x 26 x 20.5 cm); Lent by Deutsches Museum, Munich; Signed: F. Barbedienne, Fondeur

2.65 Statue of woman holding grapes (probably Swiss), 1890–1900, aluminum, 17⅜ x 7⅞ x 4⅜ in. (45 x 20 x 11 cm); Lent by Alusuisse Technology & Management Ltd., Communications TCA, Neuhausen, Switzerland

2.66 Alcoa (American, 1888–present)
Coatrack, 1924, cast aluminum and paint, H. 71½ x w. 17½ in. (181.6 x 44.5 cm); Carnegie Museum of Art, Gift of Mellon Bank, 1999.25.1

2.67 Alcoa (American, 1888–present)
Table, 1924, cast aluminum and paint, 24 x 24³⁄₁₆ x 12 in. (61 x 61.4 x 30.5 cm); Carnegie Museum of Art, Gift of Mellon Bank, 1999.25.2

2.68 Chair (American), ca. 1968, cast aluminum, 30 x 20½ x 18 in. (76.2 x 52.1 x 45.7 cm); Lent by Historical Society of Western Pennsylvania, Gift of Aluminum Company of America

2.69 George Steedman (American, 1871–1940)
Garden chair, ca. 1930, aluminum, 49½ x 26 x 24 in. (125.7 x 66 x 61 cm); Lent by Casa del Herrero, Montecito, California

2.70 Model of facade of Théâtre Porte Saint-Martin, Paris, 1891, aluminum, brass, glass, and wood, 34¼ x 35⅝ x 12⅝ in. (87 x 90 x 32 cm); Lent by Musée d'Orsay, Paris, Gift of the Société des Amis du Musée d'Orsay, 1990

2.71 Otto Wagner, designer (Austrian, 1841–1918); J & J Kohn, manufacturer (Austrian, 1868–1923)
Die Zeit chair, 1902, beech, aluminum, tape, and fabric, 30¹⁵⁄₁₆ x 22⁷⁄₁₆ x 20⁷⁄₁₆ in. (78.6 x 57 x 51.9 cm); Carnegie Museum of Art, Purchase: Gift of Trustees, Fellows and Women's Committee in honor of Phillip M. Johnston, the Henry J. Heinz II Director of the Carnegie Museum of Art, 1996.40

2.72 Otto Wagner (Austrian, 1841–1918)
Full-scale replica of 1902 Die Zeit facade, 1984–85, aluminum and glass, H. 177 x w. 130 in. (449.6 x 330.2 cm); Lent by Historisches Museum der Stadt Wien

## Aluminum and the Modernist Ideal

3.1 Dudley Talcott (American, 1899–1986)
Gate, ca. 1929–32, aluminum, 80 x 60 x 1 in. (203.2 x 152.4 x 2.5 cm); Lent by the Estate of Dudley Talcott

3.2 Jean Prouvé (French, 1901–1984)
Panel from the facade of the Fédération du Bâtiment (Building Trades Federation) headquarters, Paris, 1949, aluminum, glass, and other materials, 114⅛ x 56⅝ x 5½ in. (290 x 144 x 14 cm); Carnegie Museum of Art, DA2000.19

3.3 Overdoor decoration at Alcoa's Aluminum Research Laboratories, New Kensington, Pennsylvania, 1929, aluminum, H. 36 x w. 112 in. (91.4 x 284.4 cm); Lent by Alcoa Inc.

3.4 Wendell August Forge (American, 1923–present)
Main gate panels at Alcoa's Aluminum Research Laboratories, New Kensington, Pennsylvania, 1929, aluminum, H. 87 x w. 58 in. (221 x 147.3 cm); Lent by Alcoa Inc.

**3.5** Interior grill at Alcoa's Aluminum Research Laboratories, New Kensington, Pennsylvania, 1929, aluminum, H. 42 x W. 36 in. (106.7 x 91.4 cm); Lent by Alcoa Inc.

**3.6** SEOS Displays Limited, designer and manufacturer (British, 1984–present); Starnet International Corporation, designer and manufacturer (American, 1986–present)
Design test section for the Chrysler Technology Center Styling Dome, ca. 1992, aluminum, 120 x 240 x 30 in. (304.8 x 609.6 x 76.2 cm); Lent by SEOS Displays Limited

**3.7** A. Lawrence Kocher, architect (American, 1885–1969); Albert Frey, architect (American, b. Switzerland, 1903–1998)
Aluminaire House, 1931, model, made by University Art Museum staff, 1992; foamcore, clear plastic, white plastic, silver foil, and wood; model: 14¾ x 11¼ x 30½ in. (37.5 x 28.6 x 77.5 cm); base: 1¼ x 24½ x 30½ in. (3.2 x 62.2 x 77.5 cm); Architecture and Design Collection, University Art Museum, University of California, Santa Barbara

**3.8** R. Buckminster Fuller, designer (American, 1895–1993); Beech Aviation, Inc., now owned by Raytheon Company, manufacturer (American, 1932–present)
Carlins from the Wichita (Dymaxion) House, 1947, aluminum, each L. 126 in. (320.04 cm); Lent by Henry Ford Museum and Greenfield Village, Dearborn, Michigan

**3.9** Otto Wagner (Austrian, 1841–1918)
Lamp LHG 1576 from the Austrian Postal Savings Bank, 1904–06, aluminum and other materials, H. 7 x DIAM. 45½ in. (18 x 115.5 cm); Lent by Austrian Postal Savings Bank, Vienna

**3.10** Jacques Le Chevallier (French, 1896–1987)
Desk lamp, ca. 1927–30, aluminum and Bakelite, H. 10 x DIAM. 9 in. (25.4 x 22.9 cm); Carnegie Museum of Art, Women's Committee Acquisition Fund, 1998.45.1

**3.11** Jacques Le Chevallier (French, 1896–1987)
Desk lamp, ca. 1927–30, aluminum and Bakelite, 16½ x 7½ x 11 in. (41.9 x 19 x 27.9 cm); Carnegie Museum of Art, Women's Committee Acquisition Fund, 1998.45.2

**3.12** Jacques Le Chevallier (French, 1896–1987)
Chistera lamp, ca. 1929, aluminum, 12¼ x 5⅛ x 14½ in. (31.1 x 13 x 36.8 cm); Carnegie Museum of Art, Women's Committee Acquisition Fund, 2000.10

**3.13** Jacques Le Chevallier (French, 1896–1987)
Desk lamp, ca. 1927–30, aluminum and Bakelite; base: 16 3/16 x 11¼ x 7½ in. (41.1 x 28.6 x 19.1 cm); Lent by The Minneapolis Institute of Arts, Gift of the Norwest Corporation, Minneapolis

**3.14** Walter Dorwin Teague, designer (American, 1883–1960); Polaroid, manufacturer (American, 1937–present)
Desk lamp, ca. 1934, Bakelite and brushed aluminum, 12¼ x 11½ x 10¼ in. (31.1 x 29.2 x 26 cm); Lent by The Minneapolis Institute of Arts, Gift of the Norwest Corporation, Minneapolis

**3.15** Attributed to Walter von Nessen, designer (American, b. Germany, 1899–1943); Pattyn Products Co., manufacturer (American, active first half of 20th century)
Table lamp, ca. 1935, aluminum, Bakelite, and glass, H. 20 x DIAM. 8 in. (50.8 x 20.3 cm); Carnegie Museum of Art, Martha Mack Lewis Fund, 1999.5.1

**3.16** Artlite Company (British, active 1930s)
Illuminated table, ca. 1935, aluminum and glass, H. 21 x DIAM. 24 in. (53.3 x 61 cm); Lent from a private collection

**3.17** Cabaret table (German), ca. 1935, cast, incised, and formed sheet aluminum, glass, and other materials, H. 26¼ x DIAM. 12 in. (66.7 x 30.5 cm); Lent by Ruth and Leonard Perfido

**3.18** Jean Prouvé (French, 1901–1984)
Wall light, 1951–52, aluminum and glass, 14⅞ x 111 1/16 x 6½ in. (37.8 x 282.1 x 16.5 cm); Carnegie Museum of Art, Women's Committee Acquisition Fund, 1999.18.1

**3.19** Margaret Bourke-White (American, 1904–1971)
Photograph of Airship Akron, set in frame made from airship parts, 1931, duralumin and silver gelatin print, H. 21¼ x W. 27⅜ in. (54.2 x 69.5 cm); Lent by James and Kathleen Manwaring

**3.20** Girder from "Shenandoah" zeppelin ZRI (American), ca. 1922, duralumin and other materials, 10 x 20 x 8 in. (25.4 x 50.8 x 20.3 cm); Lent by National Air and Space Museum, Smithsonian Institution

**3.21** Curtiss-Wright Airplane Co. (American, 1929–49)
Wing rib from Robin airplane, ca. 1930, aluminum and other materials, 8½ x 71½ x 1 in. (21.6 x 181.6 x 2.5 cm); Lent by National Air and Space Museum, Smithsonian Institution

**3.22** Douglas Aircraft Company, later renamed McDonnell-Douglas Aircraft, manufacturer (American, 1920–present)
Air screw blade for the DC-3, designed in 1934, produced in 1943; aluminum, 47¼ x 12½ x 6⅝ in. (120 x 32 x 17 cm); Lent by Jean Plateau—Institut pour l'Histoire de l'Aluminium Collection, Paris

**3.23** Hugo Junkers (German, 1859–1935)
Suitcase, ca. 1919–20, duralumin, 6 x 23⅝ x 15¾ in. (15 x 60 x 40 cm); Lent by Deutsches Museum, Munich

**3.24** Warren McArthur, designer (American, 1885–1961); Warren McArthur Corporation, manufacturer (American, 1930–48)
Radio operator seat, model #219, for American Airlines DC-4, ca. 1938–48, aluminum and leather, 38 x 20¾ x 20 in. (96.5 x 52.7 x 50.8 cm); Carnegie Museum of Art, Helen Johnston Acquisition Fund, 1998.44

**3.25** Zero Halliburton (American, 1938–present)
SE3 Slimline, ca. 1995, aluminum, polyester, and neoprene, 13¾ x 16½ x 3 in. (34.9 x 41.9 x 7.6 cm); Lent from a private collection

**3.26** Eagle (American), ca. 1930s, aluminum, H. 34½ x W. 19¼ in. (87.6 x 48.9 cm); Lent by John P. Axelrod, Boston

**3.27** Pierce-Arrow car engine (American), ca. 1925, aluminum and other materials, 49 x 64 x 24 in. (124.5 x 162.6 x 61 cm); Lent by Historical Society of Western Pennsylvania, Gift of Aluminum Company of America

**3.28** Rolls-Royce (British, 1906–present)
Silver Ghost, 1921, polished aluminum and other materials, L. 132 x W. 80 in. (335.3 x 203.2 cm); Lent by Mr. and Mrs. DeNean Stafford III

**3.29** Marcel Guiguet et Cie (French, 1929–ca. 1940)
MGC N3A, 1932, aluminum, steel, and other materials; Lent by Dominique Buisson

**3.30** Arlen Ness, designer (American, b. 1939); Arlen Ness Enterprises, manufacturer (American, 1970–present)
Smoothness motorcycle, 1998, hand-formed aluminum, billet machined aluminum, and other materials, 28 x 24 x 108 in. (71.1 x 61 x 274.2 cm); Lent by Arlen Ness Enterprises

**3.31** Paul Cret, designer (American, b. France, 1876–1945); American Flyer Manufacturing Corporation, later owned by A.C. Gilbert Company and Lionel Trains, Inc., manufacturer (American, 1907–present)
Burlington Zephyr model train, 1934, aluminum, brass, plastic, steel, and acetate, 9 x 7 x 30½ in. (22.9 x 17.8 x 77.5 cm); Lent by Mitchell Wolfson Jr. Collection, The Wolfsonian—Florida International University, Miami Beach

**3.32** Paul Cret, designer (American, b. France, 1876–1945); John F. Harbeson, designer (American, 1888–1986)
Wall ornament from Santa Fe Railroad dining car, ca. 1940, aluminum and paint, 27 x 20½ x 1 in. (68.6 x 52.1 x 2.5 cm); Lent by Mitchell Wolfson Jr. Collection, The Wolfsonian—Florida International University, Miami Beach

**3.33** Paul Cret, designer (American, b. France, 1876–1945); John F. Harbeson, designer (American, 1888–1986)
Wall ornament from Santa Fe Railroad dining car, ca. 1940, aluminum and paint, 28 x 17½ x 1 in. (71.1 x 44.5 x 2.5 cm); Lent by Mitchell Wolfson Jr. Collection, The Wolfsonian—Florida International University, Miami Beach

**3.34** Airstream Inc. (American, 1932–present)
Bambi trailer, 1962, aluminum and other materials, 102 x 81 x 192 in. (259.1 x 205.7 x 487.7 cm); Lent by Airstream Inc.

**3.35** Isamu Noguchi (American, b. Japan, 1904–1988)
*Miss Expanding Universe*, 1932, aluminum, 40⅞ x 34⅞ x 10 in. (103.8 x 88.6 x 25.4 cm); Lent by The Toledo Museum of Art, Museum Purchase

**3.36** Eileen Gray (Irish, 1878–1976)
E.1027 cupboard, 1923–28, aluminum, wood, cork, glass, and metal, 66¾ x 21¼ x 6⅜ in. (169.5 x 54 x 16.2 cm); Lent by Musée des Arts Décoratifs, Paris

**3.37** Marcel Breuer, designer (American, b. Hungary, 1902–1981); Embru-Werke AG, manufacturer (Swiss, 1904–present)
Side chair, model no. 301, 1932, aluminum and bent plywood, 29 x 17 x 18½ in. (73.7 x 43.2 x 47.0 cm); Lent by Mitchell Wolfson Jr. Collection, The Wolfsonian—Florida International University, Miami Beach

**3.38** Richard Neutra (American, 1892–1970)
Hors d'oeuvres tray, ca. 1930, anodized aluminum, 2¼ x 27⅞ x 16¼ in. (5.7 x 70.8 x 47 cm); Carnegie Museum of Art, Second Century Acquisition Fund, 1998.3a–g

**3.39** Frederick Kiesler (American, 1890–1965)
Nesting coffee table, 1935–38, cast aluminum;
(1) 9½ x 34 x 25 in. (24 x 86.5 x 63.5 cm);
(2) 9½ x 22 x 16¼ in. (24 x 56 x 41 cm);
Lent by Mrs. Frederick Kiesler

**3.40** Donald Deskey, designer (American, 1894–1989); Deskey-Vollmer, manufacturer (American, 1927–31)
Table, ca. 1930, aluminum, enameled metal, and wood, 16⅛ x 24 x 24 in. (41 x 61 x 61 cm); Cooper-Hewitt National Design Museum, Smithsonian Institution/Art Resource, New York, The Decorative Arts Association Acquisitions Fund and General Museum Funds, 1993–111–4

**3.41** Alfonso Iannelli (American, b. Italy, 1888–1965)
Candlestick, 1926, brushed aluminum, H. 36 x DIAM. 8 in. (91.4 x 20.3 cm); Lent by The Minneapolis Institute of Arts, Gift of the Norwest Corporation, Minneapolis

**3.42** Warren McArthur (American, 1885–1961)
Armchair, 1932, lacquered steel tubing, lacquered wood, and upholstery, 33 x 22 x 20 in. (83.8 x 55.9 x 50.8 cm); Carnegie Museum of Art, Women's Committee Acquisition Fund, 1999.18.2

**3.43** Warren McArthur (American, 1885–1961)
Rainbow Back chair, 1934–35, anodized aluminum and upholstery, 36½ x 18 x 19 in. (92.7 x 45.7 x 48.3 cm); Carnegie Museum of Art, Purchase: Gift of Mrs. Louise Buck and Mrs. George H. Love, 1999.11

**3.44** Frank Lloyd Wright, designer (American, 1867–1959); Steelcase Inc., manufacturer (American, 1912–present)

Armchair for Johnson Wax Administration Building, Racine, Wisconsin, designed 1937, manufactured 1938, enameled steel, walnut, and upholstery, 34½ x 23½ x 19½ in. (87.6 x 59.7 x 49.3 cm); High Museum of Art, Atlanta, Georgia, Purchase with funds from the Decorative Arts Endowment, 1984.371.2

**3.45** Frank Lloyd Wright, designer (American, 1867–1959); Warren McArthur Corporation, manufacturer (American, 1930–48)
Armchair, prototype, for Johnson Wax Administration Building, Racine, Wisconsin, designed 1937, manufactured 1938, aluminum, leather, and cloth, 32 x 23½ x 18 in. (81.3 x 59.7 x 45.7 cm); Lent by Mitchell Wolfson Jr. Collection, The Wolfsonian—Florida International University, Miami Beach

**3.46** Gerrit Rietveld, designer (Dutch, 1888–1964); possibly Gerard van de Groenekan, maker (Dutch, 1924–1994); possibly Wim Rietveld, maker (Dutch, 1924–1985)
Armchair, prototype, 1942, aluminum sheet, 27⅝ x 28¼ x 24 in. (70 x 72 x 61 cm); Lent by Stedelijk Museum, Amsterdam

**3.47** Scott Burton (American, 1939–1989)
Chair, 1980–81, lacquered aluminum, 30 x 70 x 23½ in. (76.2 x 177.8 x 59.7 cm); Lent by The Art Institute of Chicago, Gift of Lannon Foundation, 1997.135

**3.48** Frank Lloyd Wright, designer (American, 1867–1959); Blue Stem Foundry, manufacturer (American)
Secretarial chair for H. C. Price Company Tower, ca. 1956, cast aluminum, rubber, and upholstery, 33½ x 18½ x 21 in. (85.1 x 47 x 53.3 cm); High Museum of Art, Atlanta, Georgia, Purchase with funds from the Decorative Arts Endowment, upholstery gift of Martha Cade, 1998.88

**3.49** Charlotte Perriand, designer (French, 1903–1999); Jean Prouvé, designer (French, 1901–1984)
Bibliothèque Mexique, 1952, aluminum, steel, and wood, 63 x 72 x 12 in. (160 x 182.9 x 30.5 cm); Carnegie Museum of Art, Berdan Memorial Trust Fund, 1998.33

**3.50** Banania, maker (French, 1914–present)
Box, ca. 1943, aluminum and paint, 4 x 3 x 3 in. (10 x 7.8 x 7.8 cm); Lent by Jean Plateau—Institut pour l'Histoire de l'Aluminium Collection, Paris

**3.51** René Lalique (French, 1860–1945)
Four circular disks (prototypes for boxes), 1922 or earlier, stamped aluminum and lacquer, each DIAM. ⅜ in. (1 cm); Carnegie Museum of Art, Edgar L. Levenson Fund, 1997.67.1–.4

**3.52** René Lalique (French, 1860–1945)
Two circular disks (prototypes for boxes), 1922 or earlier, stamped aluminum, each DIAM. 6⅛ in. (15.6 cm); Carnegie Museum of Art, Edgar L. Levenson Fund, 1997.67.5–.6

**3.53** René Lalique (French, 1860–1945)
Prototype for box, 1922 or earlier, stamped aluminum and lacquer, H. 1¾ x DIAM. 5 in. (4.5 x 12.7 cm); Carnegie Museum of Art, DuPuy Fund, 1998.14.1a–b

**3.54** René Lalique (French, 1860–1945)
Prototype for box, 1922 or earlier, stamped aluminum and lacquer, H. 1 x DIAM. 1¾ in. (2.5 x 4.5 cm); Carnegie Museum of Art, DuPuy Fund, 1998.14.2a–b

**3.55** René Lalique, designer (French, 1860–1945); Roger & Gallet, manufacturer (French, 1862–present)
Three cosmetics boxes, 1922, stamped aluminum and lacquer, each H. ⅝–1⅛ x DIAM. 3 in. (1.6–2.9 x 7.6 cm); Carnegie Museum of Art, DuPuy Fund, 1998.14.3a,b–.5a,b

**3.56** Gibbs, manufacturer (British, then French, 1912–present)
Toothpaste container, ca. 1910, aluminum, ⅞ x 2 x 2 in. (2 x 5 x 5 cm); toothpaste container, ca. 1930, aluminum, ⅔ x 2¼ x 2¼ in. (1.7 x 5.7 x 5.7 cm); toothpaste container, ca. 1940, plastic, ⅘ x 2⅝ x 2⅝ in. (2 x 6.7 x 6.7 cm); toothpaste container, ca. 1950, aluminum and paint, ⅝ x 2½ x 2½ in. (1.7 x 6.3 x 6.3 cm); Lent by Jean Plateau—Institut pour l'Histoire de l'Aluminium Collection, Paris

**3.57** Edward Cook & Co., retailer (British)
Shaving cream container, ca. 1910, aluminum and paint, 3⅜ x 1½ x 1½ in. (8.6 x 4 x 4 cm); Lent by Jean Plateau—Institut pour l'Histoire de l'Aluminium Collection, Paris

**3.58** ERMA, manufacturer (French, active first half of 20th century)
Manual vacuum cleaner, ca. 1930, aluminum, wood, steel, yellow metal, rubber, and linen, 50 x 7½ x 9½ in. (127 x 19 x 24 cm); Lent by Jean Plateau—Institut pour l'Histoire de l'Aluminium Collection, Paris; Marked: ERMA and BSGDG (breveté sans garantie du gouvernement)

**3.59** Lurelle Guild, designer (American, 1898–1986); Electrolux L.L.C., manufacturer (American, 1924–present)
Electrolux, model XXX, 1937, chrome-plated, polished, and enameled steel, cast aluminum, vinyl, and rubber, 8½ x 23 x 7¾ in. (21.6 x 58.4 x 19.7 cm); Lent by Kristine M. Schmidt in loving memory of Frank and Helen Schmidt

**3.60** Sears, Roebuck & Co. (American, 1893–present)
Imperial Kenmore vacuum cleaner, 1930, aluminum, metal, iron, rubber, and cloth, 47 x 12½ x 13 in. (119.4 x 31.8 x 33 cm); Lent by Mitchell Wolfson Jr. Collection, The Wolfsonian—Florida International University, Miami Beach

**3.61** Cadillac (French, active in mid-20th century)
Vacuum cleaner, ca. 1950, aluminum, steel, rubber, and linen, 48⅜ x 10¼ x 7⅞ in. (123 x 26 x 20 cm); Lent by Jean Plateau—Institut pour l'Histoire de l'Aluminium Collection, Paris; Marked: Aspirateur Luxe Cadillac (France) / Distribué par Cadillac (Distribution) / 79, Champs Elysées, Paris 8è, ELY 9503

**3.62** Lurelle Guild, designer (American, 1898–1986); Aluminum Cooking Utensil Company, later renamed Mirro Aluminum Company, manufacturer (American, 1901–present)
Wear-Ever coffeepot, model no. 5052, 1932, aluminum, wood, and paint, 11 x 9¾ x 4½ in. (27.9 x 24.8 x 11.4 cm); Lent by Mitchell Wolfson Jr. Collection, The Wolfsonian—Florida International University, Miami Beach

**3.63** John Gordon Rideout, designer (American, d. 1951); Harold L. van Doren, designer (American, 1895–1957); Wagner Ware Manufacturing Company, now owned by General Housewares Corporation, manufacturer (American, 1891–present)
Magnalite teakettle, ca. 1940, aluminum and nickel alloy and lacquered wood, 7¼ x 9½ x 9 in. (18.4 x 23.5 x 22.9 cm); Lent by Mitchell Wolfson Jr. Collection, The Wolfsonian—Florida International University, Miami Beach

**3.64** Sears, Roebuck & Co. (American, 1893–present)
Maid of Honor teakettle, ca. 1930, aluminum and plastic, 7¼ x 9¼ x 8½ in. (18.4 x 23.5 x 21.6 cm); Lent by Mitchell Wolfson Jr. Collection, The Wolfsonian—Florida International University, Miami Beach

**3.65** Wagner Ware Manufacturing Company, now owned by General Housewares Corporation (American, 1891–present)
Grand Prize teakettle, 1916, aluminum and wood, 9 x 11 x 13 in. (22.9 x 27.9 x 33 cm); Lent by Mitchell Wolfson Jr. Collection, The Wolfsonian—Florida International University, Miami Beach

**3.66** Alfred Edward Burrage, designer (British); Burrage & Boyd, now owned by Staffordshire Holloware Ltd., manufacturer (British, 1932–present)
Picquot Ware K3 kettle, ca. 1950; designed 1937, manufactured 1939–40 and 1948–present; polished die-cast aluminum-magnesium alloy and sycamore, 6⅜ x 9⅝ x 8¼ in. (17 x 25 x 21 cm); Lent by Jean Plateau—Institut pour l'Histoire de l'Aluminium Collection, Paris; Marked: Picquot Ware / K3 / regd design / made in England

**3.67** Aluminum Goods Manufacturing Company, later renamed Mirro Aluminum Company (American, 1895–present)
Mirro teapot, ca. 1935, aluminum, H. 6½ (16.5 cm); Lent by Richard and Jane Nylander; Inscribed: MIRRO / The Finest Aluminum / 1526M / 1½ QT / for Better Homes Club Plan

**3.68** Bialetti Industrie (Italian, early 1950s–present)
OMG Mini Express coffeemaker, ca. 1980, cast aluminum alloy, plastic, and metal, 6 x 3 x 3⅓ in. (15.2 x 7.6 x 8.5 cm); Lent by Jean Plateau—Institut pour l'Histoire de l'Aluminium Collection, Paris; Marked: mini express / 1 tazza / OMG, design L. BIALETTI

**3.69** Bialetti Industrie (Italian, early 1950s–present)
OMG Nuova Mignon coffeemaker, ca. 1980, cast aluminum alloy, plastic, colored aluminum sheet, and metal, 4½ x 4¼ x 5½ in. (11.3 x 11 x 13.9 cm); Lent by Jean Plateau—Institut pour l'Histoire de l'Aluminium Collection, Paris; Marked: Nuova Mignon / 2 Tazze / OMG / Brevetti Bialetti

**3.70** Alfonso Bialetti, designer (Italian, 1888–1970); Bialetti Industrie, manufacturer (Italian, early 1950s–present)
Three-cup and twelve-cup Moka Express coffeemakers, 1999; designed 1930, manufactured 1933–present; cast aluminum; three-cup: 6½ x 6 x 3½ in. (16.5 x 15.2 x 8.9 cm); twelve-cup: 11 x 8¾ x 5½ in. (27.9 x 22.2 x 14 cm); Lent from a private collection

**3.71** Touring Club de France, retailer (French, 1890–1983)
Coffeemaker, ca. 1950–60, aluminum, 7⅞ x 6⅛ x 5⅝ in. (20 x 15.5 x 14.5 cm); Lent by Jean Plateau—Institut pour l'Histoire de l'Aluminium Collection, Paris; Marked and stamped: Emblem of Touring Club de France; Cafetière—camping du Touring Club de France / Modèle et marque déposés

**3.72** Cake stand (British), 1932, anodized aluminum, Bakelite, enamel, and other metal, H. 36¾ x DIAM. 11¼ in. (93.3 x 28.6 cm); Lent by Mitchell Wolfson Jr. Collection, The Wolfsonian—Florida International University, Miami Beach

**3.73** Burrage & Boyd, now owned by Staffordshire Holloware Ltd., manufacturer (British, 1932–present)
Picquot Ware tray with teapot, coffeepot, milk jug, and sugar bowl, 1999; designed 1960, manufactured 1960–present; polished die-cast aluminum-magnesium alloy and sycamore; tray: 1½ x 15¾ x 10¾ in. (3.8 x 40 x 26.7 cm); teapot: H. 6½ x w. 10 in. (16.5 x 25.4 cm); coffeepot: H. 7¾ x w. 7½ in. (19.7 x 19.1 cm); milk jug: H. 4 x w. 5½ in. (10.2 x 14 cm); sugar bowl: H. 3½ x w. 3¾ in. (8.9 x 9.5 cm); Carnegie Museum of Art, Gift of Paul Reeves, 1999.59.1–.5

**3.74** Russel Wright (American, 1904–1976)
Bun warmer, ca. 1932, spun aluminum, reed, and maple, H. 9 x w. 9 in. (22.9 x 22.9 cm); Lent by Charles Biddle and Eileen O'Hara

**3.75** Russel Wright (American, 1904–1976)
Tidbit stand, ca. 1932, spun aluminum and rattan, H. 12½ x w. 8½ in. (31.8 x 21.6 cm); Lent from a private collection, Pittsburgh

**3.76** Russel Wright (American, 1904–1976)
Lemonade pitcher, ca. 1932, spun aluminum and walnut, H. 10 x w. 10½ in. (25.4 x 26.7 cm); Lent from a private collection, Pittsburgh

**3.77** Russel Wright (American, 1904–1976)
Beverage set, ca. 1932, spun aluminum, H. 11½ x w. 16½ in. (29.2 x 41.9 cm); Lent from a private collection, Pittsburgh

**3.78** Russel Wright (American, 1904–1976)
Lidded container with handle, ca. 1932, spun aluminum and wicker handle, H. 18 x w. 10 in. (45.7 x 25.4 cm); Lent from a private collection, Pittsburgh

**3.79** Russel Wright (American, 1904–1976)
Coffee urn, ca. 1935, spun aluminum and walnut, 16 x 13 x 8¼ in. (41 x 33 x 21 cm); Lent by Brooklyn Museum of Art, Gift of Paul F. Walter, 1994.16a–c

**3.80** Lavenas, manufacturer (French, 1830–1986); L. Rozay, chaser (French, active mid-20th century)
Vase, ca. 1930, spun, embossed, and chased pewter sheet, 8¼ x 6 x 6 in. (21 x 15 x 15 cm); Lent by Jean Plateau—Institut pour l'Histoire de l'Aluminium Collection, Paris; Signed: L. Rozay

**3.81** Lavenas, manufacturer (French, 1830–1986); O. Lefebvre, chaser (French, active mid-20th century)
Vase, ca. 1945, spun, embossed, and chased aluminum sheet and lead, 10¼ x 7¼ x 7⅞ in. (26 x 18.6 x 19.8 cm); Lent by Jean Plateau—Institut pour l'Histoire de l'Aluminium Collection, Paris; Signed: O. Lefebvre

**3.82** Lavenas, manufacturer (French, 1830–1986); Irman, chaser (French, active mid-20th century)
Vase, ca. 1945, spun, embossed, and chased aluminum sheet, 10⅝ x 6 x 6 in. (27 x 15 x 15 cm); Lent by Jean Plateau—Institut pour l'Histoire de l'Aluminium Collection, Paris; Signed: Irman

**3.83** Wendell August Forge (American, 1923–present)
Desk set, 1930, aluminum, desk pad: L. 26 x w. 21½ in. (66 x 54.6 cm); letter box: 1¾ x 10 x 13½ in. (4.5 x 25.4 x 34.3 cm); letter opener: L. 9½ x w. 1 in. (24.1 x 2.5 cm); inkwell: 3¼ x 3¼ x 3¼ in. (8.4 x 8.3 x 8.3 cm); blotter: 2¾ x 6¾ x 1½ in. (7 x 17.1 x 3.8 cm); Lent by Mitchell Wolfson Jr. Collection, The Wolfsonian—Florida International University, Miami Beach

**3.84** Arthur Armour (American, 1908–1988)
Butlers tray with zodiac symbols, ca. 1940, aluminum, H. 17¼ x DIAM. 19¼ in. (43.8 x 48.9 cm); Lent by Mitchell Wolfson Jr. Collection, The Wolfsonian—Florida International University, Miami Beach

**3.85** Lurelle Guild, designer (American, 1898–1986); Kensington Inc., manufacturer (American, 1934–65)
Kensington Ware Stratford compote, 1934, aluminum and plastic, H. 6 x DIAM. 13½ in. (15.2 x 34.3 cm); Carnegie Museum of Art, Gift of Christopher Monkhouse and Sarah Nichols, 92.192

**3.86** Samuel C. Brickley, designer (American); Kensington Inc., manufacturer (American, 1934–65)
Pair of Kensington Ware Stratford candlesticks, 1939, aluminum and plastic, each H. 3½ x DIAM. 4 in. (8.9 x 10.2 cm); Carnegie Museum of Art, Gift of East End Galleries, 1998.72.1.1–.2

**3.87** Trivet Manufacturing Co. (American)
Thermette hot lunch box, ca. 1943, aluminum, leather, plastic, and fabric, 9¼ x 10½ x 5¼ in. (23.5 x 26.7 x 13.3 cm); Lent by Mitchell Wolfson Jr. Collection, The Wolfsonian—Florida International University, Miami Beach

**3.88** Relax-it Massage Company (American)
Relax-it massager, model 700, date unknown, aluminum, rubber, steel, copper, and other materials, 6½ x 7¾ x 3⅝ in. (16.5 x 19.7 x 9.2 cm); Lent by Mitchell Wolfson Jr. Collection, The Wolfsonian—Florida International University, Miami Beach

**3.89** Reynolds Metals Company (American, 1913–present)
*How to Make Money at Home with Master Metal Colorfoil*, 1929, paper, L. 11 x W. 9 in. (27.9 x 22.9 cm); Lent from a private collection

**3.90** L'Aluminium Français (French, 1911–83)
Two copies of *L'Aluminium dans le ménage: autrefois, aujourd'hui* (Aluminum in the Household: Yesterday and Today), ca. 1920, paper, each L. 24 x W. 19 in. (61 x 48.3 cm); Lent by Mitchell Wolfson Jr. Collection, The Wolfsonian—Florida International University, Miami Beach, and Collection Jean Plateau—Institut pour l'Histoire de l'Aluminium Collection, Paris

**3.91** Aluminum Goods Manufacturing Company, later renamed Mirro Aluminum Company (American, 1895–present)
*Food Surprises from the Mirro Test Kitchen*, 1925, paper, L. 4½ x W. 5⅞ in. (11.4 x 14.9 cm); Lent from a private collection

**3.92** The Aluminum Ladder Company (American)
*Aluminum Ladders for Every Requirement* 3 (October 1, 1939), paper, L. 9¼ x W. 6 in. (23.4 x 15.3 cm); Lent by National Museum of American History, Smithsonian Institution

**3.93** Club Aluminum Utensil Company, later known as Club Products Company Division of Standard International Corporation (American, active 1923–73)
*The Recipe Book for Club Aluminum Ware with Personal Service*, 1926, paper, L. 5⅝ x W. 8½ in. (14.3 x 21.6 cm); Carnegie Museum of Art, DA99.89

**3.94** Harriet Lyle Veazie, artist (American); American Way Merchandising Program, retailer (American, 1940–42)
Horse from rodeo group, ca. 1940, aluminum, 4 x 3½ x 1 in. (10.2 x 8.9 x 2.5 cm); Lent by Andrew VanStyn

**3.95** Egmont C. Arens, designer (American, 1889–1966); Theodore Brookhart, designer (American, 1898–1942); Hobart Manufacturing Co., manufacturer (American, 1897–present)
Streamliner meat slicer, model no. 410, designed 1940, manufactured 1944–85, aluminum, steel, and rubber, 13½ x 21 x 15 in. (34.3 x 53.3 x 38.1 cm); Lent by Mitchell Wolfson Jr. Collection, The Wolfsonian—Florida International University, Miami Beach

## Competition and Conflict

**4.1** André Waterkeyn, engineer; André Polak and Jean Polak, architects
Model of Atomium for 1958 Brussels World's Fair, ca. 1956–57, aluminum, steel, and wood, 24⅞ x 22⅞ x 22⅞ in. (63 x 58 x 58 cm); Lent by Erik and Petra Hesmerg, Netherlands

**4.2** Set of twelve knives (French), 1900, aluminum and steel, each 9½ x ⅞ x ¼ in. (24.2 x 2 x 0.9 cm); Lent by Jean Plateau—Institut pour l'Histoire de l'Aluminium Collection, Paris; Stamped: Exposition Universelle, Paris, 1900 / Acier fondu garanti

**4.3** Set of playing cards from the Buffalo Exposition (American), 1901, aluminum, 2½ x 3½ x 6¼ in. (6.4 x 8.9 x 15.9 cm); Lent by Mitchell Wolfson Jr. Collection, The Wolfsonian—Florida International University, Miami Beach

**4.4** Pierre Alexandre Morlon, engraver (French, 1878–1951); Commissioned by L'Aluminium Français (French, 1911–83)
Medal, 1937, aluminum, 2¼ x 2¼ x ¼ in. (6 x 6 x 0.5 cm); Lent by Jean Plateau—Institut pour l'Histoire de l'Aluminium Collection, Paris; Inscription: EXPOSITION INTERNATIONALE ART ET TECHNIQUE PARIS / 1937 / AVIONS / AUTOS / CYCLES / CHEMINS DE FER / ÉLECTRICITÉ / INDUSTRIES CHIMIQUES ET ALIMENTAIRES / ARCHITECTURE / DECORATION

**4.5** L'Aluminium Français, publisher (French, 1911–83)
*L'Aluminium à l'Exposition, Paris 1937* (visitors' guide to aluminum at the 1937 Paris Exposition), 1937, paper, H. 7¾ x L. 5 in. (19.6 x 12.6 cm); Lent by Jean Plateau—Institut pour l'Histoire de l'Aluminium Collection, Paris

**4.6** Les Ateliers d'Impression et de Cartonnages, designer and printer (French)
Presentation book for New York World's Fair, 1939, aluminum foil, gilded aluminum, paper, and board; 12 x 9⅝ x ⅞ in. (30.5 x 25 x 2 cm); Lent by Jean Plateau—Institut pour l'Histoire de l'Aluminium Collection, Paris

**4.7** Everlast Metal Products Corporation (American, active 1937–51)
Souvenir tray from the New York World's Fair, 1939, aluminum, DIAM. 11¾ in. (29.8 cm); Lent by Mitchell Wolfson Jr. Collection, The Wolfsonian—Florida International University, Miami Beach

**4.8** "Golden Temple of Jehol" souvenir plaque from the Century of Progress International Exposition, Chicago, 1934, stamped and painted aluminum, acetate, polychromed printed paper, and plywood, 4¾ x 6 15/16 x ⅜ in. (12.1 x 17.6 x 1 cm); Lent by Mitchell Wolfson Jr. Collection, The Wolfsonian—Florida International University, Miami Beach

**4.9** A. P. Company (American)
Cocktail shaker, 1934, enameled aluminum, H. 11 x DIAM. 4 in. (27.9 x 10.2 cm); Lent by Mitchell Wolfson Jr. Collection, The Wolfsonian—Florida International University, Miami Beach

**4.10** Albert Speer (German, 1905–1981)
Door grille from the German Pavilion, Paris Exposition, 1937, anodized aluminum, 33¼ x 33 x 2¾ in. (84.5 x 83.8 x 7 cm); Lent by Mitchell Wolfson Jr. Collection, The Wolfsonian—Florida International University, Miami Beach

**4.11** Hans Coray, designer (Swiss, 1906–1991); Blattmann Metallwarenfabrik AG, manufacturer (Swiss, 1838–present)
Two Landi chairs, pre-1962; designed 1938, manufactured 1939–present; molded, heat-treated, and stained aluminum and rubber; 30 x 21¼ x 24½ in. (76.2 x 54 x 62.2 cm); Carnegie Museum of Art, Second Century Acquisition Fund, 2000.9.1.1–.2

**4.12** Rene Chambellan (French, 1893–1955)
Plaque, 1934, aluminum, 20 x 20 x ⅜ in. (50.8 x 50.8 x 1 cm); Lent by John P. Axelrod, Boston; Signed: Rene P. Chambellan / sculptor / 1934/12/30

**4.13** Herbert Bayer, designer (American, b. Austria, 1900–1985); Wohnbedarf, retailer (Swiss, 1931–present)
*das federnde Aluminium-Möbel* (Springy Aluminum Furniture), 1933, paper, H. 5⅞ x W. 8¼ in. (14.9 x 21 cm); Carnegie Museum of Art, Decorative Arts Purchase Fund, 1997.63

**4.14** Marcel Breuer, designer (American, b. Hungary, 1902–1981); Embru-Werke AG, manufacturer (Swiss, 1904–present)
Chaise longue, no. 313, designed 1932, manufactured 1934; aluminum and beech, 28 15/16 x 56 3/16 x 23 7/16 in. (73.5 x 142.7 x 59.5 cm); Carnegie Museum of Art, Berdan Memorial Trust Fund, 1997.41

**4.15** Gio Ponti, designer (Italian, 1891–1979); Kardex Italiano, manufacturer (Italian, active first half of 20th century)
Chair from the Montecatini Corporation headquarters, Milan, 1938, aluminum, painted steel, and padded leatherette, 30¼ x 18½ x 18½ in. (76.8 x 47 x 47 cm); Lent by Mitchell Wolfson Jr. Collection, The Wolfsonian—Florida International University, Miami Beach

**4.16** Gio Ponti (Italian, 1891–1979)
Swiveling chair #4, prototype, from the Montecatini Corporation headquarters, Milan, ca. 1937, aluminum, 30¾ x 15¾ x 16 in. (78.1 x 40 x 40.6 cm); Carnegie Museum of Art, Women's Committee Acquisition Fund, 2000.19.2

**4.17** Gio Ponti, designer (Italian, 1891–1979); Ditta Parma Antonio e Figli, manufacturer (Italian, 1870–present)
Swiveling chair #3 from the Montecatini Corporation headquarters, Milan, 1938, aluminum, steel, and vinyl, 30⅛ x 16 x 17 in. (76.5 x 40.6 x 43.2 cm); Carnegie Museum of Art, Women's Committee Acquisition Fund, 2000.19.1

**4.18** Postcard (Italian), ca. 1927, aluminum, L. 3½ x D. 5½ in. (8.9 x 14 x 73 cm); Lent by Mitchell Wolfson Jr. Collection, The Wolfsonian—Florida International University, Miami Beach

**4.19** Clive Latimer, designer (British); Heal & Son, manufacturer (British, active 1940s)
Plymet cabinet, prototype, 1945–46, cast and sheet aluminum, sheet steel, and birch veneer, 33⅞ x 53⅛ x 15¾ in. (86 x 135 x 40 cm); Lent from a private collection

**4.20** Ernest Race, designer (British, 1913–1964); Race Furniture Ltd., manufacturer (British, 1945–present)
BA3 Chair, designed 1945, manufactured 1945–69 and 1989–present, stove-enameled cast aluminum and upholstery, 28¾ x 17½ x 16¼ in. (73 x 44.5 x 41.3 cm); Carnegie Museum of Art, Decorative Arts Purchase Fund, 1998.4

**4.21** Ludwig Mies van der Rohe, designer (German, 1886–1969); Knoll, Inc., manufacturer (American, 1938–present)
Two Barcelona chairs, 1953, designed in steel 1929, manufactured in aluminum 1953 as a special commission for Alcoa Building; aluminum and leather, each 29⅞ x 30⅛ x 29¼ in. (75.8 x 76.5 x 74.3 cm); Carnegie Museum of Art, Gift of Aluminum Company of America, 95.74.1.1 and DA95.231.1

**4.22** Harrison & Abramowitz (American, 1945–76)
Model of Alcoa Building, 1953, aluminum, wood, glass, and other materials, 54 x 30 x 19 in. (137.1 x 76.2 x 48.2 cm); Lent by Carnegie Mellon University Architecture Archive

**4.23** Harrison & Abramowitz (American, 1945–76)
Interior panels from Alcoa Building, 1953, aluminum, various sizes; Lent by Regional Enterprise Tower

**4.24** Two-piece dress with belt (American), ca. 1958, aluminum thread and rhinestones; top: L. 21 x w. 15 in. (53.3 x 38.1 cm); skirt: L. 27 in. (68.5 cm); belt: L. 30½ in. (77.5 cm); Lent by Rowena and Everett Smith, Jr.

**4.25** H. S. Crocker Co. (American)
Postcard for Kaiser Aluminum Exhibit in Tomorrowland at Disneyland, ca. 1955, paper, L. 5½ x w. 3½ in. (13.9 x 8.9 cm); Lent from a private collection

**4.26** Jean Desses (Greek, b. Egypt, 1904–1970)
Stole, 1956, Lurex, silk, and mink; Lent by Brooklyn Museum of Art, Gift of Aluminum Company of America, 57.30

**4.27** Marianne Strengell (American, b. Finland, 1909–1998)
Rug, 1956, aluminum, jute, wool, and viscose, L. 147 x w. 66 in. (373.3 x 167.6 cm); Lent by Historical Society of Western Pennsylvania, Gift of Aluminum Company of America

**4.28** Isamu Noguchi (American, b. Japan, 1904–1988)
Two Alcoa Forecast Program tables, 1957, aluminum and paint, each 14¾ x 18½ x 16 in. (37.5 x 47 x 40.6 cm); Lent by Torrence M. Hunt, Sr.

**4.29** Hellerich & Bradsby Company (American, 1884–present)
Louisville Slugger softball bat, model SB25, 2000, aluminum and rubber, L. 34 in. (86.4 cm); Carnegie Museum of Art, DA2000.28

**4.30** Easton Sports (American, 1922–present)
Redline C-Core softball bat, 2000, aluminum, graphite, and leather, L. 34 in. (86.4 cm); Carnegie Museum of Art, DA2000.29

**4.31** Hellerich & Bradsby Company (American, 1884–present)
Roberto Clemente's Louisville Slugger baseball bat, made between 1955 and 1972, wood, approx. L. 34 in. (86.4 cm.); Lent by the Pittsburgh Pirates

**4.32** Spalding Sports Worldwide Inc. (American, 1876–present)
Fred Clarke's baseball bat, made between 1900 and 1915, wood, approx. L. 34 in. (86.4 cm.); Lent by the Pittsburgh Pirates

**4.33** Easton Sports (American, 1922–present)
Softball bat, model SK10 3021, ca. 1996, aluminum and rubber, L. 30 in. (76.2 cm); Lent by Craig M. Vogel and Dé Dé Greenberg

**4.34** Wilson Sporting Goods (American, 1916–present)
T2000 tennis racket, ca. 1975, designed 1967; aluminum, nylon, and other materials, L. 26⅞ x w. 10¼ in. (68.3 x 26 cm); Lent by Craig M. Vogel and Dé Dé Greenberg

**4.35** Spalding Sports Worldwide, Inc. (American, 1876–present)
Kro-bat tennis racket and press, 1960s, wood, nylon, and other materials, 2¼ x 27 x 11¼ in. (5.7 x 68.6 x 28.6 cm) (with press); Carnegie Museum of Art, DA2000.30

**4.36** Prince Manufacturing, Inc. (American, 1970–present)
Oversize tennis racket, 1987; designed 1976, manufactured 1976–present; graphite, nylon, and leather, 1⁹⁄₁₆ x 27 x 11 in. (3.9 x 68.6 x 27.9 cm); Lent by Stephen P. Webster

**4.37** Spalding Sports Worldwide, Inc. (American, 1876–present)
Smasher tennis racket, 1971, aluminum and nylon, H. 10 x L. 26½ in. (25.4 x 67.3 cm); Lent by Spalding Sports Worldwide, Inc.

**4.38** Renard Storey, designer (American, b. 1923) Bloomingdale's, manufacturer and retailer (American, 1872–present)
Bloomingdale Jetboard, 1963, aluminum and other materials, 9 x 32 x 121 in. (22.8 x 81.3 x 307.2 cm); Lent by PowerSki International Corporation

**4.39** Bob Montgomery, inventor (American, b. 1948) Bjorn Elvin, designer (Swedish, b. 1946) PowerSki International Corporation, manufacturer (American, 1987–present)
Ignitor 2000 PowerSki Jetboard, 2000, designed 1998, fiberglass, aluminum, and other materials, 9 x 26 x 100 in. (22.8 x 66 x 255 cm); Lent by PowerSki International Corporation, designed with Pro/ENGINEER by PTC using award-winning Compaq Professional Workstations

**4.40** Skateboard (probably American), ca. 1980, aluminum sheet and rubber, 4 x 8½ x 28 in. (10.3 x 21.5 x 71 cm); Lent by Jean Plateau—Institut pour l'Histoire de l'Aluminium Collection, Paris; Marked: Max-Trax GTM

**4.41** Pair of roller skates (probably American), ca. 1930s, aluminum and other materials, each 3¾ x 3⅝ x 9½ in. (9.5 x 9.2 x 24.1 cm); Lent from a private collection

**4.42** Precor USA, a subsidiary of Illinois Toolworks, Inc. (American, 1980–present)
Amerec 610 rowing machine, 1980–84, aluminum and other materials, 8 x 48 x 28 in. (20.3 x 121.9 x 71.1 cm); Carnegie Museum of Art, DA99.52

**4.43** Cannondale (American, 1971–present)
F400 bicycle, designed ca. 1995, manufactured 1995–present; aluminum and other materials, 48 x 66 x 24 in. (121.9 x 167.6 x 61 cm); Carnegie Museum of Art, DA2000.40

**4.44** Marc Newson, designer (Australian, b. 1963); Biomega, manufacturer (Danish, 1998–present)
MN-01, Extravaganza bicycle, 2000; designed 1998, manufactured 1999–present; superplastic aluminum and other materials, 39⅜ x 23⅝ x 66⅞ in. (100 x 60 x 170 cm); Lent by Biomega, Hellerup, Denmark

**4.45** Charles Eames, designer (American, 1907–1978); Ray Eames, designer (American, 1912–1988); Herman Miller, Inc., manufacturer (American, 1923–present)
Two Aluminum Group armchairs, ca. 1965; designed 1958, manufactured 1958–present; aluminum and vinyl, 35 x 25¼ x 28 in. (88.9 x 64.1 x 71.1 cm); Carnegie Museum of Art, Gift of Kimball and Diana Nedved, 1999.41 and DA99.41.2

**4.46** Eero Saarinen, designer (American, 1910–1961); Knoll, Inc., manufacturer (American, 1938–present)
Tulip Chair, designed 1956, manufactured 1956–present; plastic, cast aluminum with fused plastic finish, and upholstery; H. 31⅝ x w. 19½ x D. 21⅜ in. (80.5 x 49.5 x 54 cm); Lent by Vitra Design Museum, Weil am Rhein, Germany

**4.47** Folding chair (American), ca. late 1960s, aluminum and nylon, 31¾ x 22⅛ x 10¼ in. (80.6 x 56.2 x 26 cm); Lent by Craig M. Vogel and Dé Dé Greenberg

**4.48** Massimo Vignelli, designer (Italian, b. 1931); Heller Designs, Inc., manufacturer (American, 1970–present) Heller Dinnerware Stacking Mugs, designed 1972, manufactured 1972–present; plastic, each H. 4¼ x W. 4¾ x DIAM. 3¼ in. (10.7 x 12 x 8.3 cm); Carnegie Museum of Art, DA99.87.1-.4, and Courtesy of retromodern.com, Atlanta, Georgia

**4.49** Heller Hostess-ware (American, 1946–ca. 1955) Colorama tumblers, sherbet dishes, and pitcher, ca. 1950, anodized aluminum, various sizes; Lent by Craig M. Vogel and Dé Dé Greenberg

**4.50** Coca-Cola Company (American, 1886–present) Group of containers, 1915–present, aluminum, glass, plastic, and steel, various sizes; Lent by Craig M. Vogel and Dé Dé Greenberg, Dave Tanner, and Carnegie Museum of Art, DA2000.31

**4.51** Audi (German, 1909–present) Audi A8 4.2 quattro space frame, manufactured 1994–present; aluminum, 45¾ x 183 x 69 in. (116.2 x 464.8 x 175.2 cm); Lent by Audi and Alcoa Automotive

**4.52** Freeman Thomas, designer (American, b. 1957); Audi, manufacturer (German, 1909–present) Audi TT Coupe filler cap, 1994, aluminum, DIAM. 6 in. (15.2 cm); Lent by Audi of America, Inc.

**4.53** J Mays (American, b. 1954) et al., designers Model of multisport pickup concept car, 1995–97, fiberglass and other materials, 19 x 41¾ x 19 in. (48.3 x 106 x 48.3 cm); Lent by Alcoa Automotive

**4.54** J Mays (American, b. 1954) et al., designers Model of multisport sedan concept car, 1995–97, fiberglass and other materials, 18 x 41½ x 18½ in. (45.7 x 105.4 x 47 cm); Lent by Alcoa Automotive

**4.55** J Mays (American, b. 1954) et al., designers Model of multisport utility concept car, 1995–97, fiberglass and other materials, 18½ x 41 x 18¾ in. (47 x 104.1 x 47.6 cm); Lent by Alcoa Automotive

**4.56** Unidentified, chaser (French, active ca. 1915); Rigaworks, manufacturer (Russian, active beginning of 20th century) Canteen, manufactured 1913, chased ca. 1915; chased aluminum, 7½ x 4¾ x 2½ in. (19 x 12 x 6.5 cm); Lent by Jean Plateau—Institut pour l'Histoire de l'Aluminium Collection, Paris; Marked: C.R. / A mes chers parents / Souvenir de ma captivité (To my dear parents: Souvenir of my captivity)

**4.57** Holy water container made from canteen (French), 1915, aluminum, 10⅝ x 5⅛ x 3 in. (27 x 13 x 7.5 cm); Lent by Jean Plateau—Institut pour l'Histoire de l'Aluminium Collection, Paris; Marked: Munster 1914–15

**4.58** Inkwell and two pens (French), ca. 1914–16, aluminum and copper; inkwell: 2⅜ x 4½ x 2½ in. (6.1 x 11.3 x 6.4 cm); pens: each 7½ x 1¼ x ¼ in. (19.2 x 3.2 x 0.6 cm); Lent by Jean Plateau—Institut pour l'Histoire de l'Aluminium Collection, Paris; Inkwell marked: 1914 and 1915; pens marked: Verdun 1916 / Marguerite and Argonne AD

**4.59** Oscar Cloquette (French) Box, ca. 1916, engraved aluminum and yellow metal, 1½ x 6½ x 3¼ in. (3.6 x 16.7 x 8.2 cm); Lent by Jean Plateau—Institut pour l'Histoire de l'Aluminium Collection, Paris; Marked: interlaced O and J / Toujours / Oscar CLOQUETTE / pendant son exil / Soltou [Germany] / 1914–1915–1916

**4.60** Louis Oppenheim (German, 1879–1936) Poster, 1916, paper, H. 18½ x W. 27¼ in. (47.5 x 69 cm); Lent by Kölnisches Stadtmuseum, Cologne

**4.61** Asipim chair (African), 20th century, wood, aluminum, and steel, 40 x 18 x 17 in. (101.6 x 45.7 x 43.2 cm); Carnegie Museum of Art, Overs African Art Fund, 1998.43

**4.62** Fon staff (African), possibly 1930s, aluminum, iron, and enamel inlay, H. 36½ in. (92.7 cm); Lent by Mr. Colin Sayer, The Collector, Cape Town, South Africa

**4.63** John Garrett (American, b. 1950) *Pod Pod #4*, 1998, reused aluminum beverage cans and hardware cloth, 13 x 19 x 19 in. (33 x 48.3 x 48.3 cm); Carnegie Museum of Art, James L. Winokur Fund, 1998.60

**4.64** Boris Bally (American, b. 1961) Two Transit chairs, 1997, reused aluminum traffic signs, each 48 x 16 x 21 in. (121.9 x 40.6 x 53.3 cm); Carnegie Museum of Art, Decorative Arts Purchase Fund, 1997.39.1.1–.2

**4.65** Hisanori Masuda, designer (Japanese, b. 1949); Kikuchi Hojudo Inc., manufacturer (Japanese, 1604–present); Chushin-Kobo, manufacturer (Japanese, 1997–present) Iquom Tableware Collection "Egg" and "Oval" jewelry boxes, 1999; designed 1992, manufactured 1992–present; sand-cast recycled aluminum and gold leaf, w. 3½–5 x L. 2½–3¾ in. (8.9–12.7 x 6.4–9.5 cm); Lent from a private collection

**4.66** Clare Graham (American, b. 1949) *Carpet of Printed Can Labels*, 1997, reused aluminum beverage cans, L. 144 x W. 144 in. (365.8 x 365.8 cm); Lent by the artist

**4.67** Clare Graham (American, b. 1949) *Serpentine Chaise Longue*, 1997, reused aluminum beverage cans, 37½ x 90 x 25 in. (95.3 x 228.6 x 63.5 cm); Lent by the artist

**4.68** Hat (Vietnamese), 1999, reused aluminum Coca-Cola cans and other materials, 4½ x 11½ x 7½ in. (11.4 x 29.2 x 19 cm); Lent by Craig M. Vogel and Dé Dé Greenberg

**4.69** Purse (Vietnamese), ca. 1998, reused Coca-Cola cans and other materials, 4 x 3¼ x 1½ in. (10.1 x 8.3 x 3.8 cm); Carnegie Museum of Art, DA98.102

**4.70** Jet (Vietnamese), ca. 2000, reused Coca-Cola cans and other materials, 4 x 12½ x 7¼ in. (10.1 x 31.8 x 18.4 cm); Carnegie Museum of Art, DA2000.32

**4.71** Bořek Šípek, designer (Czech, b. 1949); Vitra AG, manufacturer (Swiss, 1934–present) Sedlak chair, 1992, aluminum, beech, and molded polyurethane, 34¼ x 15¾ x 20 in. (87 x 40 x 51 cm); Lent by Elaine Caldwell, New York

**4.72** Philippe Starck, designer (French, b. 1949); Vitra AG, manufacturer (Swiss, 1934–present) Louis 20 armchair, 1999; designed 1991, manufactured 1992–present; polished aluminum and blown polypropylene, 32¼ x 23½ x 23 in. (84.5 x 59 x 60 cm); Carnegie Museum of Art, Purchase: Gift of Marilyn P. and Thomas J. Donnelly in honor of Phillip M. Johnston, by exchange, 2000.16

## Crossing Boundaries

**5.1** Paco Rabanne (Spanish, b. 1934) Minidress, 1969, aluminum and polished silver squares, H. 29½ x W. 24 in. (75 x 61 cm); Lent by Jean Plateau—Institut pour l'Histoire de l'Aluminium Collection, Paris

**5.2** Paco Rabanne (Spanish, b. 1934) Atomium Bruxelles Dress, 1999, aluminum and stainless steel, L. 43¼ in. (110 cm); Lent by the designer

**5.3** Paco Rabanne (Spanish, b. 1934) Triangle Dress with Headdress, 1999, aluminum and stainless steel, L. 68½ in. (175 cm); Lent by the designer

**5.4** Lamex, manufacturer (French, active second half of 20th century) Butcher's apron, ca. 1970–80, aluminum and linen, 22 x 19¼ x ⅜ in. (56 x 49 x 1 cm); Lent by Jean Plateau—Institut pour l'Histoire de l'Aluminium Collection, Paris; Marked: Lamex

**5.5** Peter Muller-Munk Associates (American, 1938–present) Concept drawing of living room with glass and aluminum, ca. 1969, paper, pencil, and ink, L. 13¼ x W. 16¹⁵⁄₁₆ in. (33.7 x 43 cm); Carnegie Museum of Art, Gift of Dirk Visser, DA97.34.2

**5.6** Kappler Safety Group (American, 1976–present)
Reflector® protective overcover, designed 1991, manufactured 1994–present; aluminum and other materials; Carnegie Museum of Art study collection, Gift of Kappler Safety Group

**5.7** Issey Miyake (Japanese, b. 1939)
Dress and hat from the Starburst series, autumn/winter collection, 1998, aluminum and cotton; Lent by Miyake Design Studio

**5.8** Nigel Coates (British, b. 1949)
Mannequin, 1993, aluminum, 56 x 14¼ x 7 in. (142 x 36 x 18 cm); Lent by The Montreal Museum of Fine Arts / Montreal Museum of Decorative Arts, The Liliane and David M. Stewart Collection

**5.9** Nigel Coates (British, b. 1949)
Mannequin, 1993, aluminum, 29¼ x 17 x 8½ in. (74 x 43 x 21.5 cm); Lent by The Montreal Museum of Fine Arts / Montreal Museum of Decorative Arts, The Liliane and David M. Stewart Collection

**5.10** Pollyanna Beeley (British, b. 1970)
Bodice, 1993, anodized aluminum, L. 14 x w. 17¾ in. (35.6 x 45 cm); Lent by the artist

**5.11** Oscar de la Renta (American, b. Dominican Republic, 1936)
Bathing suit, ca. 1967, aluminum and synthetic fabric, L. 53½ in. (135.9 cm); Lent by Historical Society of Western Pennsylvania, Gift of Aluminum Company of America

**5.12** Pollyanna Beeley (British, b. 1970)
Evening handbag, 2000, anodized aluminum, 10¼ x 5 x 3⅜ in. (26 x 12.7 x 8.6 cm); Lent from a private collection

**5.13** Handbag (German), ca. 1930, aluminum, enamel, and silk, 5¾ x 8¾ x 1¾ in. (14.6 x 22.2 x 4.4 cm); Lent by The Minneapolis Institute of Arts, Gift of the Norwest Corporation, Minneapolis

**5.14** Salvatore Ferragamo Company (Italian, 1927–present)
Shoes, 1955–56, suede, aluminum, silk, and leather, L. 8⅝ x H. 3½ in. (22 x 9 cm); Lent by Museo Salvatore Ferragamo

**5.15** Salvatore Ferragamo Company (Italian, 1927–present)
Handbag, 1996–97, anodized aluminum, nickel, and fabric, 6¾ x 7⅞ x 3⅛ in. (17 x 20 x 8 cm); Lent by Museo Salvatore Ferragamo

**5.16** Salvatore Ferragamo Company (Italian, 1927–present)
Handbag, 1996–97, anodized aluminum, nickel, and fabric, 11¾ x 7⅜ x 3⅞ in. (30 x 19 x 10 cm); Lent by Museo Salvatore Ferragamo

**5.17** Reiko Sudo, designer (Japanese, b. 1953); Nuno Corporation, manufacturer (Japanese, 1984–present)

Rusted Silver Washer, 1991, cotton, polyester, and aluminum lamé, L. 259 x w. 41 in. (658 x 104.1 cm); Carnegie Museum of Art, DA2000.34

**5.18** Junichi Arai (Japanese, b. 1932)
Lunar Stream, 1999, silk, polyester, and aluminum, L. 236¼ x w. 45¼ in. (600 x 115 cm); Carnegie Museum of Art, DA2000.35

**5.19** Junichi Arai (Japanese, b. 1932)
Blue, 1995, wool, polyester, and aluminum, L. 283½ x w. 36¼ in. (720 x 92 cm); Carnegie Museum of Art, DA2000.36

**5.20** Emeco (American, 1944–present)
1006 chair, 1999; designed 1944, manufactured 1944–present; aluminum, 34½ x 15¾ x 20 in. (87.6 x 40 x 50.8 cm); Carnegie Museum of Art, Second Century Acquisition Fund, 2000.9.2.1

**5.21** The General Fireproofing Company (American, 1902–present)
Chair, 1940s, aluminum, 33½ x 15½ x 20½ in. (85 x 39.4 x 52 cm); Lent by Kennametal, Inc.; Marked: Good Form, Made in USA, Youngstown, Ohio

**5.22** Philippe Starck, designer (French, b. 1949); Emeco, manufacturer (American, 1944–present)
Two Hudson chairs, 2000; designed 1999–2000, manufactured 2000; brushed and polished aluminum, each 34½ x 15¾ x 20 in. (87.6 x 40 x 50.8 cm); Carnegie Museum of Art permanent collection, Gift of Emeco

**5.23** Philippe Starck, designer (French, b. 1949); Driade SpA, manufacturer (Italian, ca. 1985–present)
Romantica chair, 1987, aluminum sheet and aluminum tubing, 35 x 25 x 16 in. (88.9 x 63.5 x 40.6 cm); Carnegie Museum of Art, Gift of Celia Morrissette and Keith Johnson, 1999.40

**5.24** Maarten Van Severen, designer (Belgian, b. 1949); Maarten Van Severen Meubelen, manufacturer (Belgian, 1988–present)
Low Chair, 1999; designed 1993–95, manufactured 1995–present; aluminum, 24¾ x 36¼ x 19¾ in. (63 x 92 x 50 cm); Lent from a private collection

**5.25** Philippe Starck, designer (French, b. 1949); Vitra AG, manufacturer (Swiss, 1934–present)
W. W. Stool, 1999; designed 1990, manufactured 1992–present; varnished sand-cast aluminum, 38¾ x 21 x 21¼ in. (98.4 x 53.3 x 54 cm); Carnegie Museum of Art, Edgar L. Levenson Fund, 1999.31.2

**5.26** Jorge Pensi, designer (Spanish, b. Argentina, 1946); Amat s.a., manufacturer (Spanish, 1944–present)
Toledo chair, designed 1986–88, manufactured 1989–present; epoxy-coated cast aluminum and aluminum tubing, 30 x 21¾ x 21¼ in. (76.2 x 55.3 x 54 cm); Carnegie Museum of Art, Second Century Acquisition Fund, 2000.17

**5.27** Frank Gehry, designer (American, b. Canada, 1929); Knoll, Inc., manufacturer (American, 1938–present)
FOG chair, 2000; designed 1999, manufactured 2000; aluminum and stainless steel, 31¾ x 25 x 24 in. (80.6 x 63.5 x 61 cm); Carnegie Museum of Art, Gift of Knoll, Inc., 2000.31

**5.28** Marc Newson, designer (Australian, b. 1963); Marc Newson, Ltd., manufacturer (British, 1997–present)
Orgone chair, 1999; designed 1993, manufactured 1999–present; plastic, 37¾ x 42½ x 32 in. (96 x 108 x 81 cm); Lent from a private collection

**5.29** Marc Newson, designer (Australian, b. 1963); Pod, manufacturer
Orgone chair, 2000, designed 1993, polished aluminum and paint, 37¾ x 42½ x 32 in. (96 x 108 x 81 cm); Carnegie Museum of Art permanent collection

**5.30** Ron Arad, designer (Israeli, b. 1951); Ron Arad & Associates, manufacturer (British, 1989–present)
Un-cut, 1997, anodized aluminum and stainless steel, 31⅝ x 38 x 38⅞ in. (80.3 x 96.5 x 98.7 cm); Carnegie Museum of Art, DuPuy Fund, 1997.65

**5.31** Ron Arad, designer (Israeli, b. 1951); Ron Arad & Associates, manufacturer (British, 1989–present)
Tom Vac, 1997, aluminum and stainless steel, 31½ x 28¼ x 19¾ in. (80 x 72 x 50 cm); Lent by Ron Arad & Associates

**5.32** Ron Arad, designer (Israeli, b. 1951); Vitra AG, manufacturer (Swiss, 1934–present)
Tom Vac, 1999; designed 1997, manufactured 1999–present; molded polypropylene and chrome, 29¾ x 25½ x 24 in. (75.6 x 65 x 61 cm); Carnegie Museum of Art, Edgar L. Levenson Fund, 1999.31.1

**5.33** Ali Tayar, designer (American, b. Turkey, 1959); Parallel Design Partnership, manufacturer (American, 1922–present); Wernerco, extruder of aluminum elements (American, 1922–present)
Ellen's Brackets, 2000; designed 1993, manufactured 1993–present; extruded aluminum, L. 10 x w. 1 in. (25.4 x 2.5 cm); Carnegie Museum of Art permanent collection, Gift of the designer and Parallel Design Partnership

**5.34** Ali Tayar, designer (American, b. Turkey, 1959); ICF Group, manufacturer (American, 1962–present); Wernerco, extruder of aluminum elements (American, 1922–present)
Rasamny chair, 1999, aluminum and wood, 31½ x 20 x 22 in. (80 x 50.8 x 55.9 cm); Carnegie Museum of Art, Gift of ICF Group, 2000.30

**5.35** Ali Tayar, designer (American, b. Turkey, 1959); ICF Group, manufacturer (American, 1962–present); Wernerco, extruder of aluminum elements (American, 1922–present)
Plaza screen, 1999, aluminum, H. (minimum) 72 in. x w. 72 in. (182.9 x 182.9 cm); Lent by ICF Group

**5.36** Aaron Lown (American, b. 1968)
Hi Ho Stool, prototype, 1994, vacuum-laminated fiberglass, urethane foam, machined aluminum, sand-cast aluminum, rubber, and leather, 28½ x 17½ x 17½ in. (72.4 x 44.5 x 44.5 cm); Lent by the designer

**5.37** Ron Arad, designer (Israeli, b. 1951); Kartell SpA, manufacturer (Italian, 1945–present) FPE (Fantastic Plastic Elastic) chair, designed 1997, manufactured 1999–present; plastic and extruded aluminum, 31½ x 16½ x 20¾ in. (80 x 41.9 x 52.7 cm); Carnegie Museum of Art, Purchase: Gift of Mrs. John J. Bissell, by exchange, 2000.7

**5.38** Shiro Kuramata, designer (Japanese, 1934–1991); Ishmaru Co., Ltd., manufacturer (Japanese, 1969–present) Miss Blanche, designed 1988, manufactured 1989–98; acrylic resin, paper flowers, and aluminum, 35½ x 21½ x 23½ in. (90 x 55 x 60 cm); Lent by The Montreal Museum of Fine Arts / Montreal Museum of Decorative Arts, The Liliane and David M. Stewart Collection

**5.39** Alberto Meda, designer (Italian, b. 1945); Alias Srl, manufacturer (Italian, 1979–present) Softlight chair, 1989, honeycomb aluminum, molded carbon fibers in epoxy-resin matrix, and elastic fiber, 29¼ x 15 x 20 in. (74.5 x 38 x 51 cm); Lent by Vitra Design Museum, Weil am Rhein, Germany

**5.40** Ross Lovegrove, designer (Welsh, b. 1958); Bernhardt Design, manufacturer (American, 1889–present) Go chair, 2000; designed 1999, manufactured 2000; 380 aluminum alloy and polycarbonate plastic, 30 x 23¼ x 26½ in. (76.2 x 59 x 67.3 cm); Carnegie Museum of Art permanent collection

**5.41** Werner Schmidt (Swiss, b. 1953) Alu-Falttisch, 1987, aluminum, H. 29 x DIAM. 25½ in. (73.7 x 64.8 cm); Carnegie Museum of Art, Martha Mack Lewis Fund, 1999.5.2

**5.42** Donald Judd, designer (American, 1928–1984); Janssen, manufacturer (Dutch, 1957–present) Corner chair, 2000; designed 1984, manufactured 1984–present; painted aluminum, 29½ x 19¾ x 19¾ in. (75 x 50 x 50 cm); Carnegie Museum of Art permanent collection

**5.43** Donald Judd, designer (American, 1928–1984); Janssen, manufacturer (Dutch, 1957–present) Armchair, 2000; designed 1984, manufactured 1984–present; anodized aluminum, 29½ x 19¾ x 19¾ in. (75 x 50 x 50 cm); Carnegie Museum of Art permanent collection

**5.44** Stuart Basseches, designer (American, b. 1960); Judith Hudson, designer (American, b. 1959); Biproduct, manufacturer (American, 1998–present) I-beam table, 1999, anodized aluminum, powder-coated aluminum, and acrylic, 13 x 48 x 13 in. (33 x 121.9 x 33 cm); Carnegie Museum of Art, Gift of the designers and Biproduct, 2000.23

**5.45** David Tisdale (American, b. 1956) Two pairs of servers, 1987, aluminum, ½ x 8⅞ x 2⅞ in. (1.3 x 22.5 x 7.3 cm); Lent from a private collection

**5.46** David Tisdale (American, b. 1956) Serving tray, 1984, aluminum, 1⅛ x 14 x 9 in. (2.9 x 35.6 x 22.9 cm); Lent by The Montreal Museum of Fine Arts / Montreal Museum of Decorative Arts, Gift of David Tisdale

**5.47** Afra Scarpa, designer (Italian, b. 1937); Tobia Scarpa, designer (Italian, b. 1935); San Lorenzo Srl, manufacturer (Italian, 1970–present) Three dishes, 2000; designed 1992, manufactured 1992–present; anodized and varnished aluminum, DIAM. 7⅞, 11¾, and 13 in. (20, 30, and 33 cm); Lent from a private collection

**5.48** Yoshiharu Fuwa, designer (Japanese); Authentics Artipresent GmbH, manufacturer (German, 1988–present) Three kettles, 1989, anodized aluminum and plastic, each H. 8¼ x W. 7¾ x DIAM. 7¼ in. (21 x 19.7 x 18.4 cm); Lent from a private collection and Kölnisches Stadtmuseum, Cologne

**5.49** Ryohin Keikaku Co., Ltd., manufacturer (Japanese, 1979–present); Muji, retailer (Japanese, 1983–present) Group of objects, 2000, aluminum, various sizes; Carnegie Museum of Art, DA2000.37.1-.28

**5.50** Gijs Bakker (Dutch, b. 1942) Circle bracelet, 1967, anodized aluminum, 15⅛ x 4⅛ x 2 in. (13 x 10.5 x 5 cm); Lent by the artist, Courtesy Helen Drutt: Philadelphia

**5.51** Gijs Bakker (Dutch, b. 1942) Circle bracelet, 1967, polished aluminum, 4 x 3⅛ x 2¼ in. (10 x 8 x 6 cm); Lent by the artist, Courtesy Helen Drutt: Philadelphia

**5.52** Jane Adam (British, b. 1954) Five Pod brooches, 2000, dyed and crazed anodized aluminum, gold leaf, and freshwater pearls, each L. approx. 3 in. (7.6 cm); Lent by the artist

**5.53** Jane Adam (British, b. 1954) Four bangles, 2000, dyed and crazed anodized aluminum, each approx. H. 2 x DIAM. 3 in. (5.1 x 7.6 cm); Lent by the artist

**5.54** Jane Adam (British, b. 1954) Boa necklace, 2000, dyed and crazed anodized aluminum, gold leaf, freshwater pearls, and stainless-steel wire, L. approx. 30 in. (76.2 cm); Lent by the artist

**5.55** Marcia Lewis (American, b. 1946) Stethoscope Neckpiece, ca. 1978–80, aluminum, feathers, laminated vegetable ivory, and ebony, 8 x 9 x 6 in. (20.3 x 22.9 x 15.2 cm); Carnegie Museum of Art, Gift of the artist, 1999.39.3

**5.56** Frans van Nieuwenborg, maker (Dutch, b. 1941); Martijn Wegman, maker (Dutch, b. 1955) Necklace, 1972, aluminum, polyester, and plastic, DIAM. 6¼ x D. ⅜ in. (16.1 x 1.8 cm); Cooper-Hewitt, National Design Museum, Smithsonian Institution / Art Resource, New York, Museum purchase from the Smithsonian Institution Collection Acquisitions Program and the Decorative Arts Association Acquisitions Fund, 1995–8–1

**5.57** Shiang-shin Yeh (American, b. Taiwan, 1969) Bracelet, 1997, anodized aluminum, 5½ x 5½ x 1½ in. (14 x 14 x 3.6 cm); Carnegie Museum of Art, James L. Winokur Fund, 2000.18

**5.58** Arline Fisch (American, b. 1931) Pleated Fan / Blue necklace, 1991, pleated and anodized aluminum, H. 8 x L. 12 in. (20.3 x 30.5 cm); Lent by the artist

**5.59** Arline Fisch (American, b. 1931) Necklace, 1984, pleated and anodized aluminum, W. 3 x DIAM. 14 in. (7.6 x 35.6 cm); Lent by the artist

**5.60** Arline Fisch (American, b. 1931) Springy Flower necklace, 1990, anodized aluminum, W. 10 x DIAM. 5 in. (25.4 x 12.7 cm); Lent by the artist

**5.61** Sony Electronics Inc. (Japanese, 1946–present) Two Aibo robotic dogs, 1999, aluminum, plastic, and other materials, each 10½ x 16 x 5½ in. (26.7 x 40.6 x 14 cm); Lent by Entertainment Robot America, Sony Electronics Inc.

**5.62** Carnegie Mellon University Robot Learning Laboratory (American, 1990–present) Jeeves, the Tennis Rover, 1996, aluminum, plastic, and other materials, 16⅜ x 24½ x 25½ in. (41.6 x 62.2 x 64.8 cm); Lent by Carnegie Mellon University Robot Learning Laboratory

**5.63** Karim Rashid (Canadian, b. Egypt, 1960) *Digitalia Time Capsule*, 1999, aluminum, LED display, and microcomputer, 30 x 18 x 18 in. (76.2 x 45.7 x 45.7 cm); Lent by Karim Rashid, Inc.

**5.64** Marc Newson, designer (Australian, b. 1963); Pod, manufacturer MN-01 LC1, Lockheed Lounge, 1986–88, designed 1985, riveted sheet aluminum over fiberglass and rubber, 25 x 35 x 60 in. (63.5 x 88.9 x 152.4 cm); Carnegie Museum of Art, Women's Committee Acquisition Fund, 1999.30

**5.65** Chrysler Corporation, now DaimlerChrysler Corporation (American, 1924–present) Prowler, manufactured 1996–present, aluminum and other materials, 51 x 165 x 76 in. (129.5 x 419.1 x 193 cm); Lent by DaimlerChrysler Corporation and Alcoa Automotive

# Index

## Photography Credits